CO-BKU-034

621.38152
4/90

Published by:
Semiconductor Services
Redwood City, CA
U.S.A.

Based on "Microchip Fabrication"
by P. Van Zant

copyright 1989

Preliminary Release
Fourth Printing 1990
ISBN 0-9623880-4-8

Edited by: Warren Landry
               Carol Rose

Printed in the United States of America

A

AC: see "Alternating Current"

ASIC: Application Specific Integrated Circuit

ABSOLUTE VISCOSITY: the viscosity of a liquid, measured in centipoise. Three methods are used to measure viscosity.

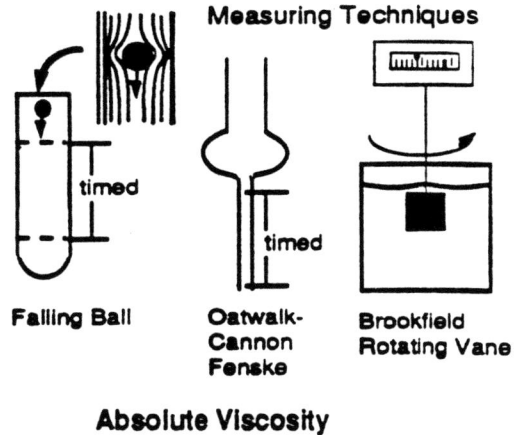

Measuring Techniques

Falling Ball

Oatwalk-
Cannon
Fenske

Brookfield
Rotating Vane

Absolute Viscosity

ACCELERATION TUBE: a high voltage tube which uses a large difference in voltage (in the 100,000 volt range) to move ions through it at increasing speeds.

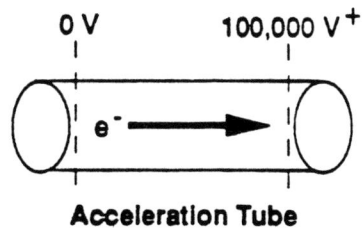

0 V   100,000 V$^+$

e$^-$

Acceleration Tube

ACCEPTOR: a dopant from Group III of the periodic table. An acceptor replaces a silicon atom in the silicon crystal structure. Because an acceptor has one less electron than silicon, it leaves a "hole" (positively charged region) in the cystal structure, making it P-type silicon.

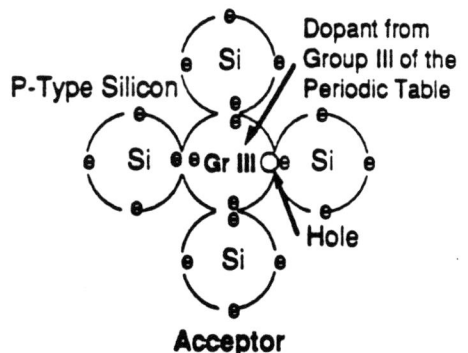

P-Type Silicon

Dopant from
Group III of the
Periodic Table

Si

Gr III

Si

Si

Si

Hole

Acceptor

ACCUMULATIVE YIELD: see "Fabrication Yield"

ACOUSTIC WAVE DEVICE: a non-silicon device used in communication systems, particularly microwave applications. It converts electromagnetic waves to acoustic waves.

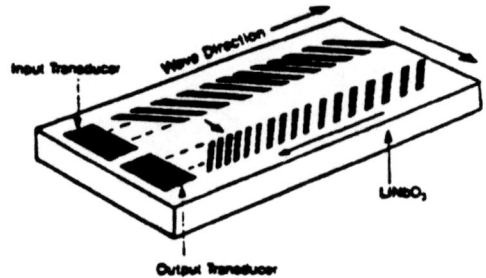

Acoustic Wave Device

ACTIVE COMPONENT: an electrical component which includes an energy source.

ADHESION: the ability of materials such as photoresist to adhere to the variety of surfaces used in semiconductor processing.

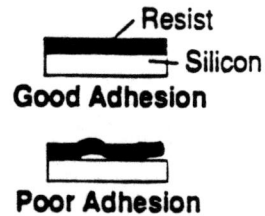

AIR CLASS NUMBER: the number of particulates of a specified size per cubic foot per minute, measured at various locations in the semiconductor manufacturing plant. Specified by Federal Specification # 209D e.g. Class 100 no more than 100 particles larger than 0.5 microns per cubic foot per minute.

Air Class Numbers

Environment	Class #	Particle Size Micron
VLSI Area	10	0.3
VLF Hood	100	0.5
Typical Fab Area	10,000	0.5
House Room	100,000	0.5
Outdoors	> 100,000	0.5

2

ALIGNER: a system which optically transfers an image from a mask to a wafer.

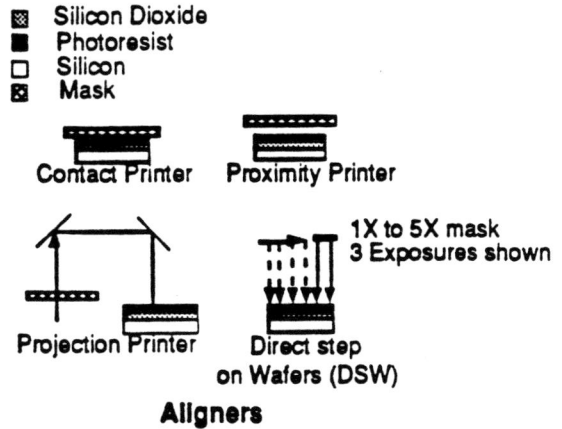

Silicon Dioxide
Photoresist
Silicon
Mask

Contact Printer   Proximity Printer

Projection Printer   Direct step
on Wafers (DSW)

1X to 5X mask
3 Exposures shown

Aligners

ALIGNMENT: refers to the positioning of a mask or reticle with respect to the wafer. For the first masking step, the wafer flat is used. In subsequent steps, alignment marks or targets are used to insure that the patterns are properly positioned with respect to those previously imaged. **Global** alignment refers to the positioning of the entire mask pattern to the wafer or to the underlying layer on the wafer. **Site- by-site** or **die -by-die** alignment refers to the positioning of the die pattern from a layer with respect to its corresponding die on the previous layer.

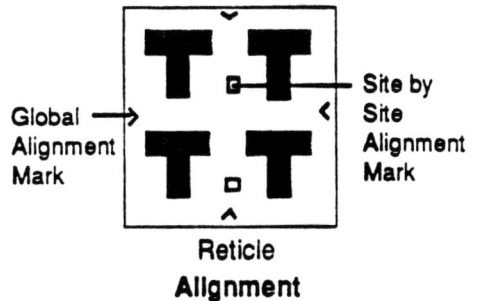

Global Alignment Mark

Site by Site Alignment Mark

Reticle
Alignment

ALIGNMENT MARKS: registration targets used to insure that the entire wafer image is positioned correctly with respect to the underlying layer (global alignment) and that die are positioned correctly with respect to the underlying die image (die-by-die or site-by-site). Alignment accuracy is determined by measuring the difference in distance (in x and y directions) in alignment targets from consecutive layers.

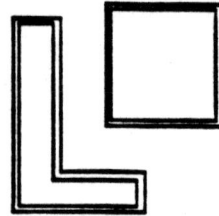

Alignment Marks

ALLOY: 1) a compound composed of two metals. 2) in semiconductor processing, the alloy step causes the interdiffusion of the semiconductor and the metallic material on top of it, forming an ohmic contact .

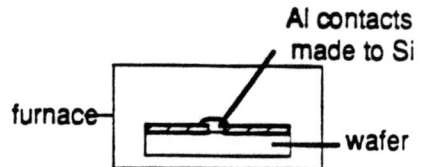

The Alloy Process

AlCu
An Alloy

ALLOY JUNCTION TRANSISTOR: an early transistor formed in germanium with junctions formed at the interface of certain metals alloyed to the germanium.

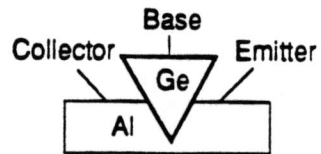

Alloy Junction Transistor

ALTERNATING CURRENT (AC): an electric current that reverses its direction regularly and continually.

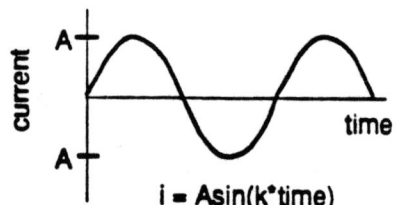

$i = A\sin(k \cdot time)$
Alternating Current

4

ALUMINUM [Al]: the metal often used in semiconductor technology to form the interconnects between devices on a chip. It can be applied by evaporation or sputtering.

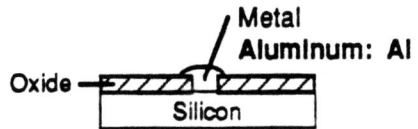

Oxide — Metal **Aluminum: Al**
Silicon

AMBIENT: surroundings: e.g. the ambient temperature in a fab is the temperature of the facility. It is assumed that materials kept in an environment where the surrounding temperature is constant will maintain that same temperature.

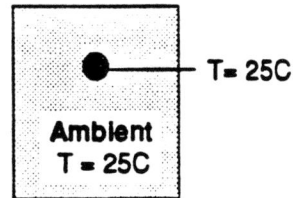

T = 25C

Ambient
T = 25C

AMORPHOUS: refers to materials with no definite arrangement of atoms. e.g. plastics

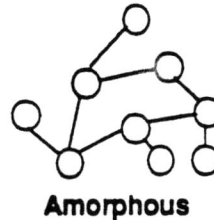

Amorphous

ANALOG: non-digital representation of data.

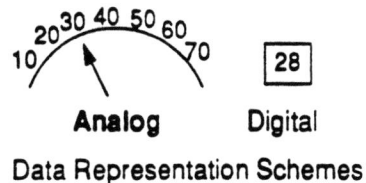

10 20 30 40 50 60 70   28

Analog   Digital

ANALOG DEVICES: devices which make use of the amplification or modification of AC signals.

Data Representation Schemes

ANGLE LAP: a method of magnifying the depth of a junction by cutting (lapping) through it at an angle away from the perpendicular.

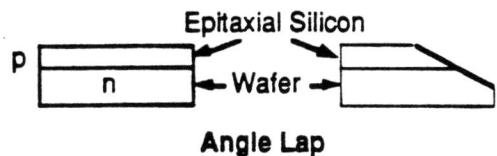

Epitaxial Silicon
p
n ← Wafer →
Angle Lap

ANGSTROM: a unit of length equal to one ten-thousandth of a micron (10^{-4} micron), 100,000,000 angstroms equal one centimeter.

$\overset{\circ}{A}$ = Angstrom = 10^{-10} m

ANISOTROPIC: an etch process that exhibits little or no undercutting resulting in features whose sides are perpendicular to the underlying layer.

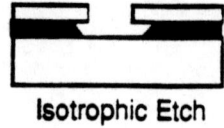

Isotrophic Etch

Anisotrophic Etch

ANNEAL: a high temperature processing step (usually the last one), designed to repair defects in the crystal structure of the wafer.

before

after **Anneal**

ANTIMONY (Sb): the N-type dopant often used as the dopant for the buried layer.

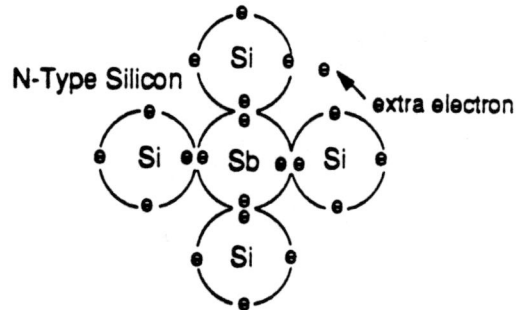

Antimony: Sb

ARRAY: an arrangement utilizing rows and columns.

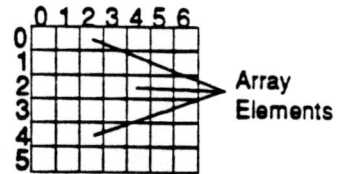

Array

ARSENIC: the N-type dopant often used for the predeposition buried layer.

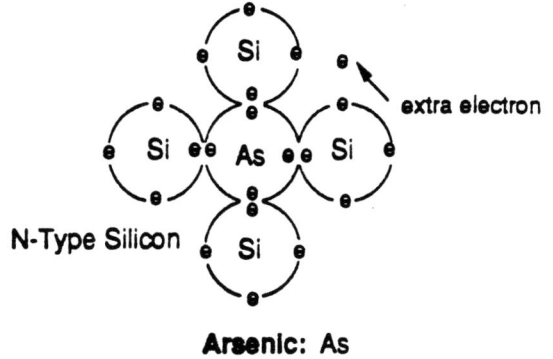

N-Type Silicon

extra electron

Arsenic: As

ASHING: a method of stripping photoresist that utilizes high energy gas.

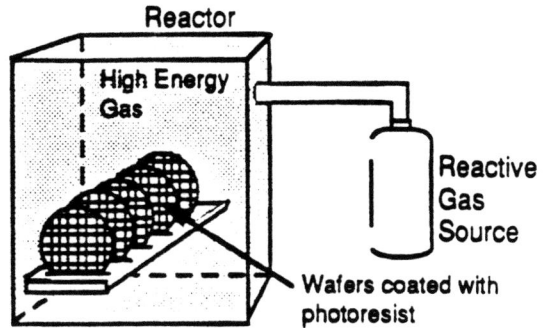

Reactor

High Energy Gas

Reactive Gas Source

Wafers coated with photoresist

Ashing

ASSEMBLY: The final stage of semiconductor manufacturing in which the wafer is divided into individual units (chips) and placed in a protective package. Steps include probing, dicing, die attach, wire bonding, encapsulation, and final test. Refered to as the "back end".

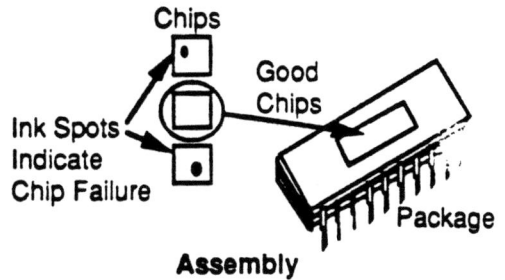

Chips

Ink Spots Indicate Chip Failure

Good Chips

Package

Assembly

ASSEMBLY YIELD: the number of die functioning after packaging as compared to the number functioning at wafer sort. Typically, this is the highest of all process yields which make up Overall Yield.

$$\frac{\text{\# Functioning Die after Packaging}}{\text{\# Functioning Die at Wafer Sort}} \times 100\% =$$

Assembly Yield

ATOMIC NUMBER: a number assigned to each element. It is equal to the number of protons (which is also equal to the number of electrons) in the atom.

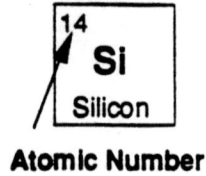

Atomic Number

ATOMIC PARTICLES: the three parts of an atom: electrons, protons and neutrons.

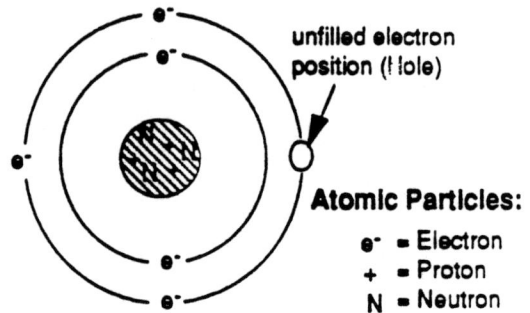

unfilled electron position (Hole)

Atomic Particles:

e^- = Electron
+ = Proton
N = Neutron

AUGER ANALYSIS: (pronounced "awe jay") a method for identification of surface materials. An E-beam is scanned across the surface of the wafer causes electrons (Auger electrons) with energies characteristic of the surface element to be released from the top several nanometers of the surface.

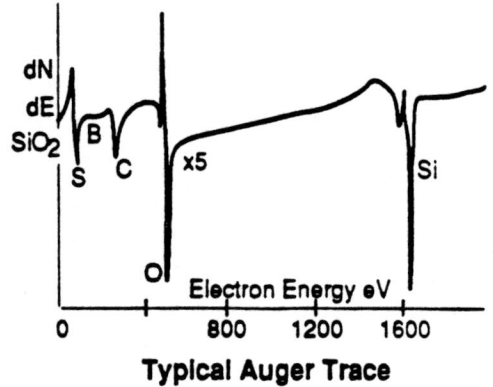

Typical Auger Trace

8

B

BOE: see "Buffered Oxide Etch"

BACKSIDE PREPARATION: an assembly process step in which the wafer is mechanically thinned to fit in a package and/or the backside is coated by evaporation with a thin metal to allow attachment of the die to the package.

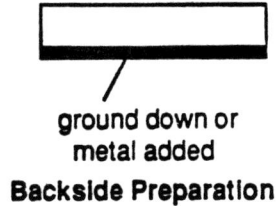

ground down or
metal added
Backside Preparation

BAR: 1) see "Chip". 2) a line of chips on a semiconductor wafer.

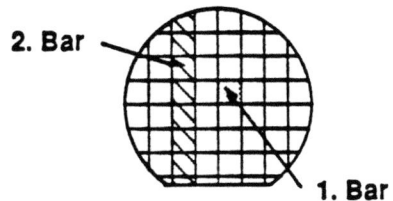

2. Bar

1. Bar

BASE: 1. the control portion of an NPN or PNP junction transistor. 2. the P-type diffusion done using boron that forms the base (1) or NPN transistors, the emitter and collector of lateral PNP transistors, and resistors.

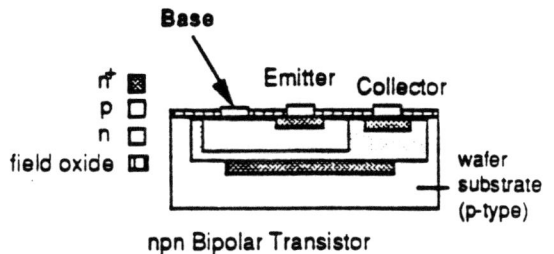

Base

n+
p
n
field oxide

Emitter   Collector

wafer
substrate
(p-type)

npn Bipolar Transistor

BATCH: a group of wafers processed together.

Batch

BEAM LEAD (for integrated circuits): a deposited metal lead, usually of gold, which projects beyond the edge of the semiconductor chip, used for both mechanical and electrical contact of the chip.

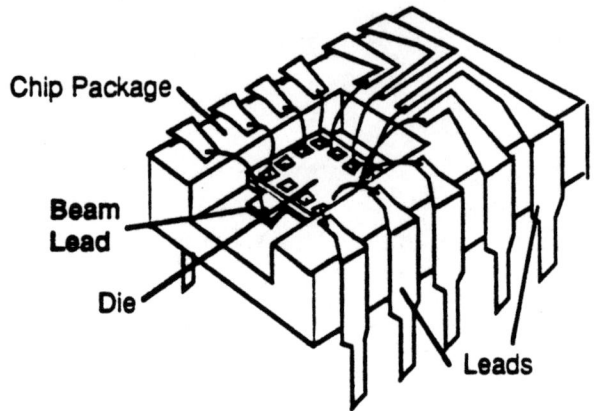

Chip Package

Beam Lead

Die

Leads

BEAM SCANNING: a method of moving an electrically charged beam across the wafer. The beam passes through a set of charged magnetic plates and is deflected onto the wafer surface. By changing the charge on the plates, the beam may be systematically moved across the wafer surface.

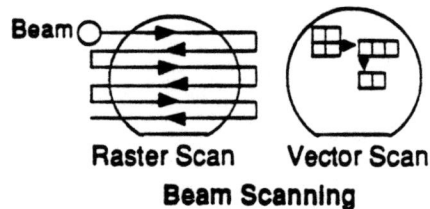

Beam

Raster Scan   Vector Scan

Beam Scanning

BERNOULI PICKUP: a wafer handling tool which uses nitrogen to create a negative pressure region which attracts the wafer to it, causing the wafer to "float" in the nitrogen making no physical contact with the handling tool.

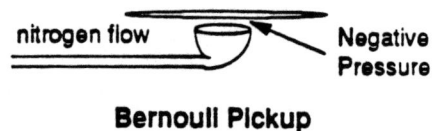

nitrogen flow

Negative Pressure

Bernouli Pickup

BINARY NOTATION: a way of representing any number using a power of two (the digits 0 and 1).

$0101 = 0·8 + 1·4 + 0·2 + 1·1 = 5$
conversion to decimal notation

Binary Notation: 101 100 001

BIPOLAR TRANSISTOR: a transistor consisting of an emitter, base and collector, whose action depends on the injection of minority carriers from the base by the collector. Sometimes called NPN or PNP transistor to emphasize its layered structure.

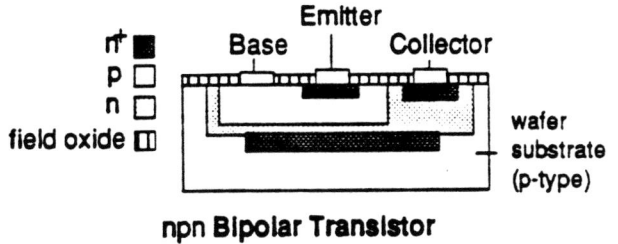

n⁺ ■
p ☐
n ☐
field oxide ▥

Emitter
Base   Collector

wafer substrate (p-type)

npn Bipolar Transistor

BIT: one digit of a binary representation, an indicator which can assume the value of 1 or 0.

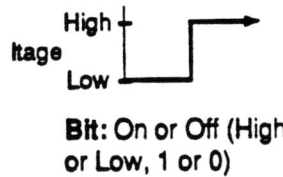

High
ltage
Low

Bit: On or Off (High or Low, 1 or 0)

BIT MAP: a translation of a pattern onto a grid in which the squares are either filled or unfilled.

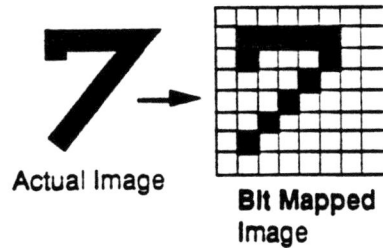

Actual Image

Bit Mapped Image

BOAT: 1. Pieces of quartz joined together to form a supporting structure for wafers during high temperature processing steps. 2. A teflon or plastic assemblage, typically made by injection molding, used to hold wafers during wet processing steps, storage, and during movement between processing steps.

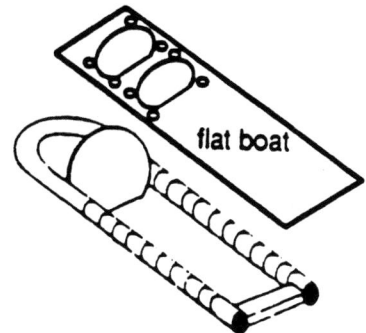

flat boat

Oxidation/Diffusion Boats

BOAT PULLER: a mechanical device used to push a boat loaded with wafers into a furnace and/or withdraw it at a fixed speed.

Boat Puller

BOHR ATOM MODEL: a model of the atom.

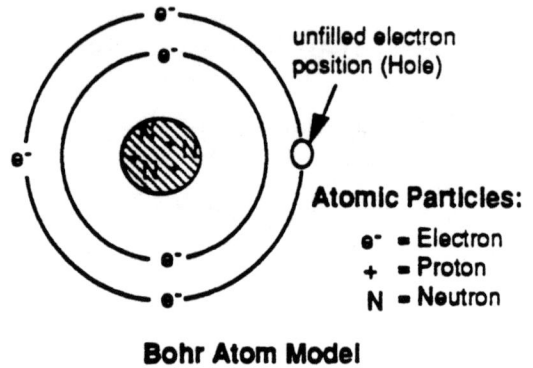

unfilled electron position (Hole)

Atomic Particles:

e^- = Electron
+ = Proton
N = Neutron

Bohr Atom Model

BONDING PAD: the relatively rectangular or square areas of metallization on a die that are the sites for electrical testing(probing) and are utilized to attach the chip to its package by wire bonding.

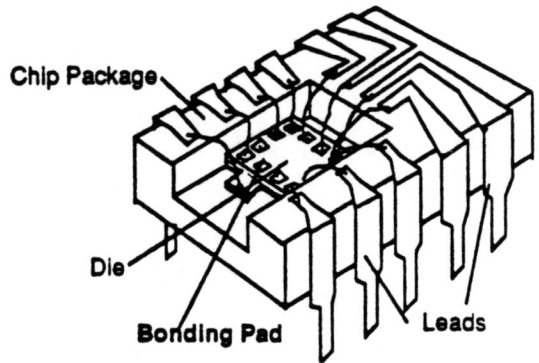

Chip Package

Die

Bonding Pad

Leads

BORON (B): the P-type dopant commonly used for the isolation and base diffusion in standard bipolar inte-grated circuit processing.

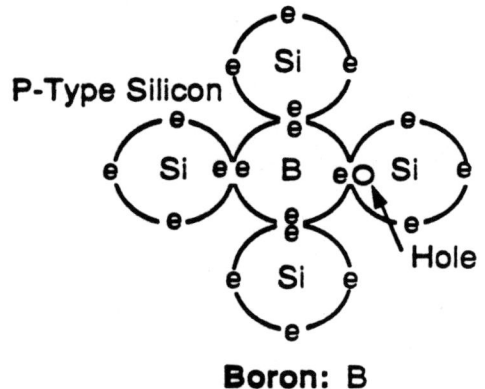

P-Type Silicon

Si

Si   B   Si

Si

Hole

Boron: B

BORON TRICHLORIDE (BCl_3): a gas that is often used as a source of boron for doping silicon.

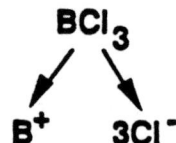

BCl_3

B^+ $3Cl^-$

BREAK: see "Pattern Defects"

BRIDGE: see "Pattern Defects"

BROWN PAD: an aluminum pad which has been exposed to etch chemicals giving it a brown, stained appearance.

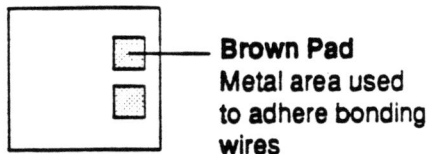

Brown Pad
Metal area used to adhere bonding wires

BUBBLER: an apparatus in which a carrier gas is transmitted through a heated liquid causing portions of the liquid to be transported with the gas, e.g. a carrier gas (nitrogen or oxygen) is "bubbled" through deionized water at 98-99 C on its way to the oxidation tube.

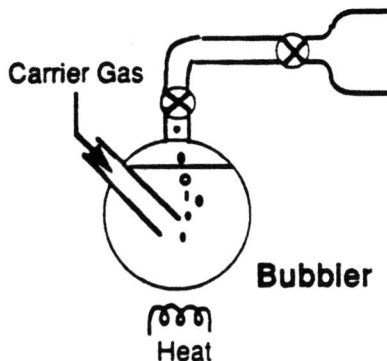

Carrier Gas

Bubbler

Heat

BUFFER: an additive that prevents the rapid change in the chemical activity of an acid or base solution by keeping the number of ions capable of reacting essentially constant even as the solution is used.

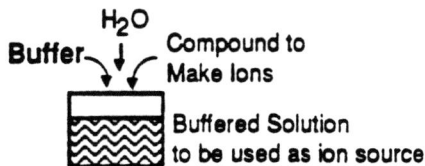

H_2O
Buffer — Compound to Make Ions
Buffered Solution to be used as ion source

BUFFERED OXIDE ETCH: a mix of hydrogen fluoride (HF) and ammonium fluoride (NH_4F) used to promote oxide etching at a slow, controlled rate.

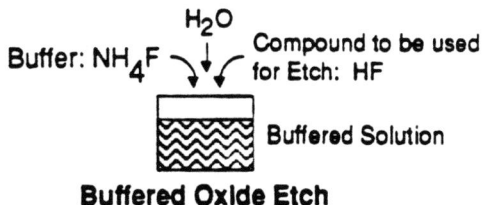

H_2O
Buffer: NH_4F — Compound to be used for Etch: HF
Buffered Solution
Buffered Oxide Etch

BUNNY SUIT: a head-to-foot overall-type garment worn by fab personnel, usually worn in Class 100 or cleaner work areas

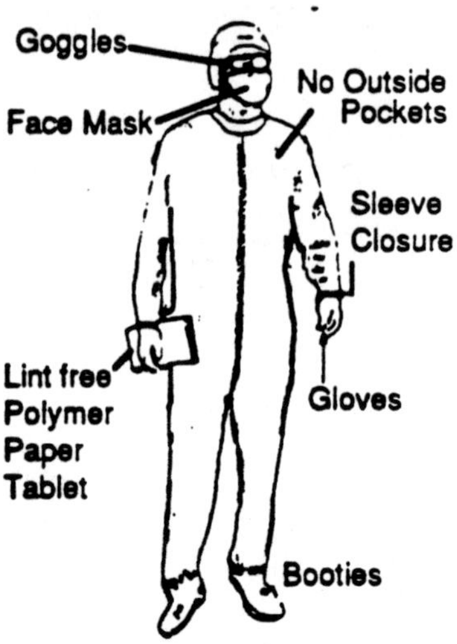

Goggles
Face Mask
No Outside Pockets
Sleeve Closure
Lint free Polymer Paper Tablet
Gloves
Booties

Bunny Suit

BURIED LAYER: the N$^+$ diffusion in a P-type substrate completed prior to growing an epitaxial layer. The buried layer provides a low resistance path for current flowing in a device. Common buried layer dopants are antimony and arsenic.

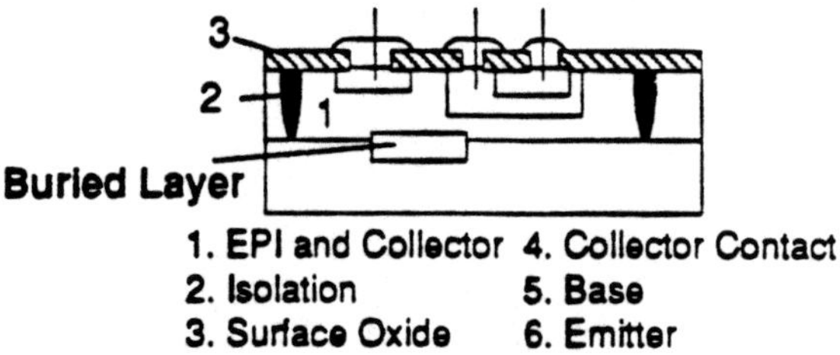

Buried Layer

1. EPI and Collector   4. Collector Contact
2. Isolation   5. Base
3. Surface Oxide   6. Emitter

Bipolar Transistor

C

CAD: see "Computer Aided Design".

CCD: see "Charge Coupled Device".

CMOS: see "Complementary Field-Effect Transistor".

CPU: see "Central Processing Unit".

CRT: see "Cathode Ray Tube".

C/V PLOT: see "Capacitance/Voltage Plot".

CVD: see "Chemical Vapor Deposition".

COMPUTER AIDED DESIGN (CAD): layout and design of integrated circuits through the use of application specific computer systems.

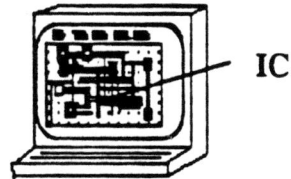

Computer Aided Design

CAN: a metal package with an array of leads extending through the base.

Can

CAPACITANCE: electrical charge storage capability.

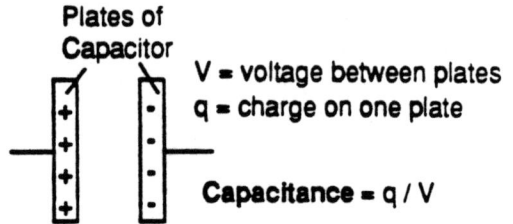

Plates of Capacitor

V = voltage between plates
q = charge on one plate

Capacitance = q / V

CAPACITANCE/VOLTAGE PLOT (C/V Plot): a plot that provides information on the amount of mobile ionic contamination present in an oxide.

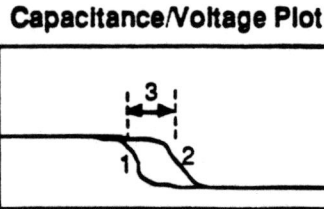

Capacitance/Voltage Plot

1. Original Plot
2. Plot as voltage increase
3. Voltage Shift or Drift

CAPACITOR: a discrete device which stores electrical charge on two conductors separated by a dielectric.

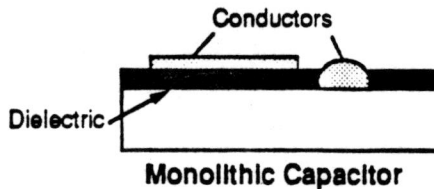

Conductors

Dielectric

Monolithic Capacitor

CAPTIVE SEMICONDUCTOR PRODUCER: a wafer manufacturer whose entire production is used by a specific company; usually by its own company.

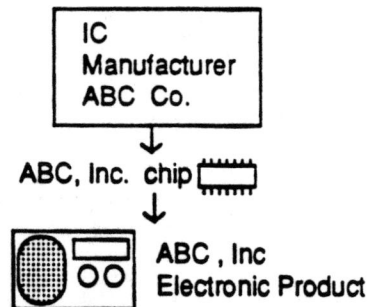

IC Manufacturer ABC Co.

ABC, Inc. chip

ABC, Inc Electronic Product

Captive Semiconductor Producer: ABC Co.

16

CARRIER GAS: an inert gas which will transport atoms or molecules of a desired substance to a reaction chamber.

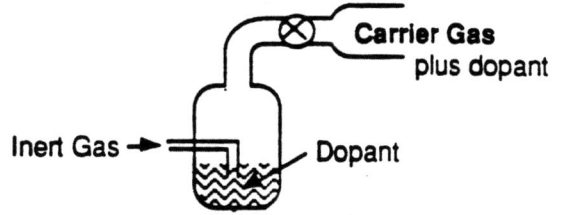

Inert Gas → Dopant

Carrier Gas plus dopant

CATHODE RAY TUBE (CRT): typically refers to a display screen which is made up of cathode ray tubes.

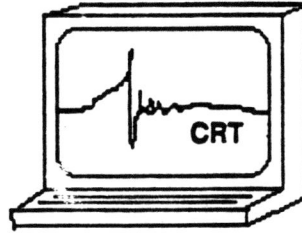

CRT

CENTISTOKES: units used to measure viscosity.

Centistokes = Centipoise / Density

CENTRAL PROCESSING UNIT (CPU):. The central control unit of a computer, usually a microprocessor.

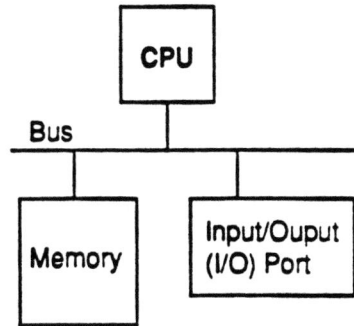

CPU

Bus

Memory

Input/Ouput (I/O) Port

CHANNEL: a thin region of a semiconductor that supports conduction. A channel may occur at a surface or in the bulk and is essential for the operation of MOSFETs and SIGFETs. In cases where channels are not part of the circuit design, their presence may indicate contamination problems or incomplete isolation.

Channel: Accidental

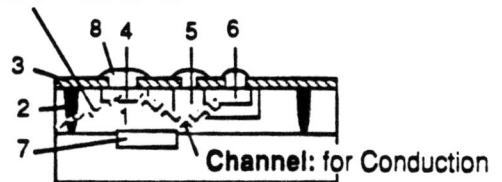

Channel: for Conduction

1. EPI and Collector
2. Isolation
3. Surface Oxide
4. Collector Contact
5. Base
6. Emitter
7. Buried Layer
8. Metallization

Bipolar Transistor

CHANNELING: a phenomenon in which an ion beam will penetrate into the crystal planes of the wafer.

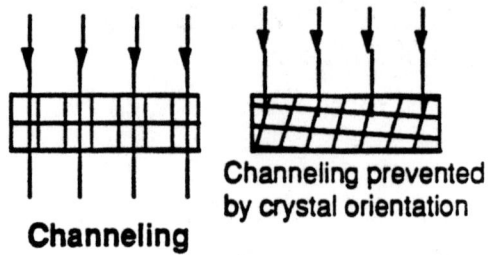

Channeling

Channeling prevented by crystal orientation

CHARGE CARRIER: a carrier of electrical charge within the crystal of a solid-state device, such as an electron or hole.

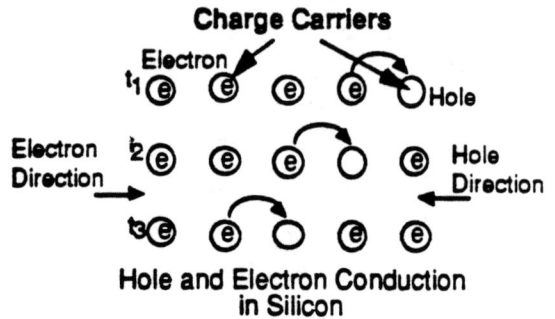

Charge Carriers

Electron

Hole

Electron Direction

Hole Direction

Hole and Electron Conduction in Silicon

CHARGE COUPLED DEVICE (CCD): a semiconductor device whose action depends on the storage of electric charge within a semiconductor by an insulated electrode on its surface.

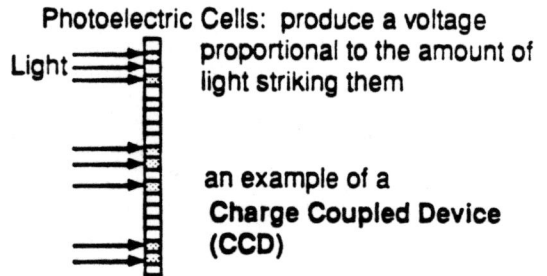

Photoelectric Cells: produce a voltage proportional to the amount of light striking them

Light

an example of a **Charge Coupled Device (CCD)**

CHEMICAL ETCHING: selective removal of material by means of chemical reactants. The precision of the etch is controlled by the temperature of the etchant, the time of contact, and the composition of the etchant.

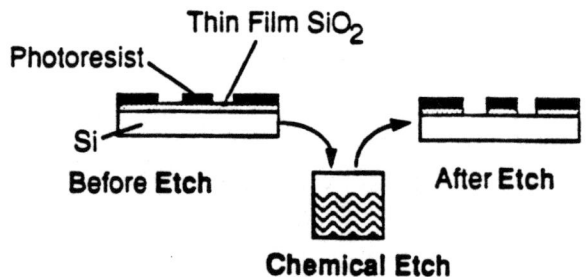

Photoresist

Thin Film SiO_2

Si

Before Etch

Chemical Etch

After Etch

18

**CHEMICAL VAPOR DEPO-
   SITION (CVD):** a
   method for depositing
   some of the layers which
   function as dielectrics,
   conductors, or semiconductors. A chemical
   containing atoms of the
   material to be deposited
   reacts with another
   chemical, liberating the
   desired material which
   deposits on the wafer
   surface while by-prod-
   ucts of the reaction are
   removed from the reac-
   tion chamber.

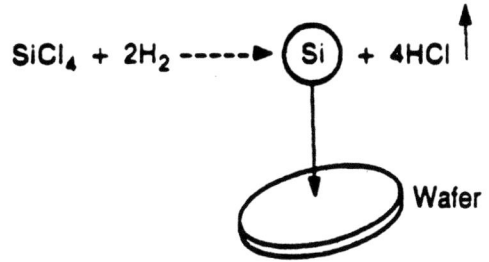

$$SiCl_4 + 2H_2 \dashrightarrow \text{Si} + 4HCl \uparrow$$

Wafer

**Chemical Vapor Deposition
of Silicon from Silicon Tetrachloride**

CHIP: (also called a die or
   device) one of the individual integrated circuits
   or discrete devices on a
   wafer. Term refers to die
   which have been cut
   from the wafer.

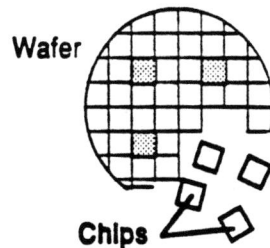

Wafer

Chips

CHROME: a metal often
   used in mask fabrication
   to form the opaque layer
   in which the circuit pattern is generated.

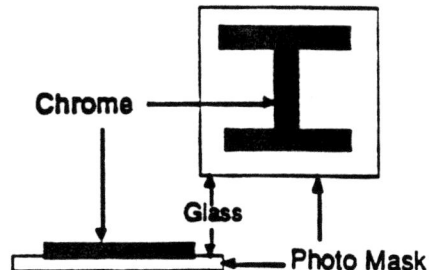

Chrome

Glass

Photo Mask

CIRCUIT BOARD: see "Printed Circuit Board"

CIRCUIT LAYOUT: the calculation of the physical device dimensions required to produce the desired electrical parameters. Vertical dimensions determine CVD and doping thickness specifications. Horizontal dimensions determine the wafer pattern dimensions and are the basis for a scaled drawing of the finished circuit (composite drawing).

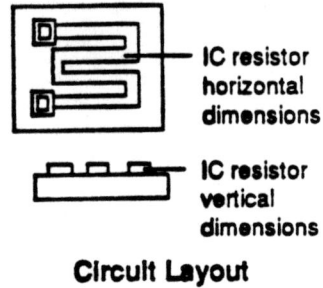

IC resistor horizontal dimensions

IC resistor vertical dimensions

Circuit Layout

CLASS NUMBER: see "Air Class Number".

Air Shower

Gowning Area

Factory

Access to Maintain Equipment from outside

Fab Area

CLEAN ROOM: a manufacturing area in which air cleanliness is controlled in order to limit the number of contaminants to which the product is exposed.

Pass Through for Chemicals

Top View of Clean Room

CLEAR FIELD MASK: a mask on which the pattern is defined by the opaque areas.

Clear Field Mask

COLLECTOR: one of the three regions of the bipolar type of transistor along with the emitter and base.

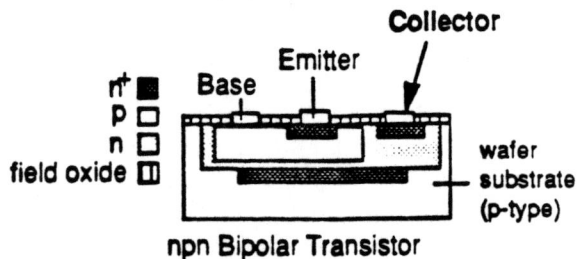

Collector

Emitter

Base

n⁺
p
n
field oxide

wafer substrate (p-type)

npn Bipolar Transistor

20

COLLIMATED LIGHT: light comprised of parallel rays used for "gross" visual inspection of surfaces.

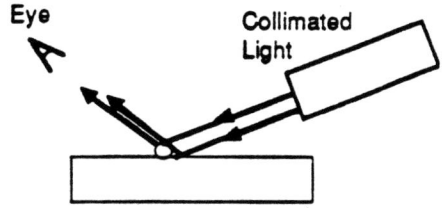

Eye   Collimated Light

Collimated Light Inspection

COMPLIMENTARY FIELD-EFFECT TRANSISTOR (CMOS): N and P channel MOS transistors on the same chip.

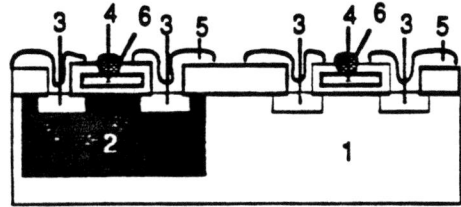

COMPONENT: an electical device such as a resisitor, fuse, capacitor, diode, or integrated circuit.

1. Wafer   4. Gate
2. P Well   5. Gate Oxide
3. Source/Drain   6. Metallization

CMOS Structure

COMPOSITE DRAWING: a scaled drawing of the finished circuit.

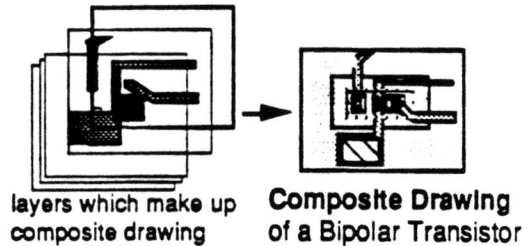

layers which make up composite drawing

Composite Drawing of a Bipolar Transistor

COMPOUND SEMICONDUCTOR: Semiconducting material formed from elements from Group III and Group V of the Periodic Table: e.g. GaAs (Gallium Arsenide).

Compound Semiconductor

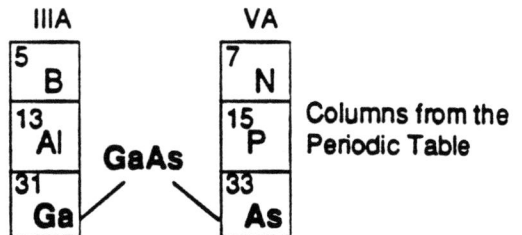

IIIA

| 5 B |
| 13 Al |
| 31 Ga |

VA

| 7 N |
| 15 P |
| 33 As |

GaAs

Columns from the Periodic Table

CONCENTRATION VS. DEPTH GRAPH: a graph which gives information on the concentration of a given dopant as a function of depth of penetration into the wafer surface.

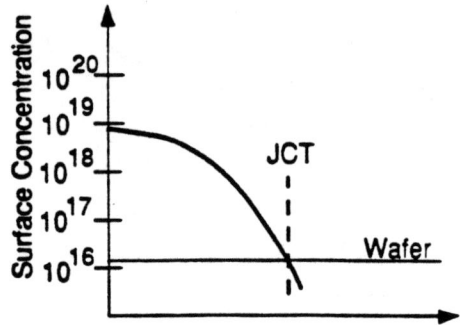

Concentration vs. Depth Graph

CONDUCTION: one of the three methods of heat transfer (conduction, convection, and radiation). For this type of heat transfer, the source and the object to be heated are in physical contact. Examples of conduction heating are: electric burners, hot plates and immersion heaters.

Conduction

CONDUCTIVITY: the ability of materials to conduct electricity Units for conductance are "siemens". Often measured indirectly by its inverse (resistance) in units of "ohms".

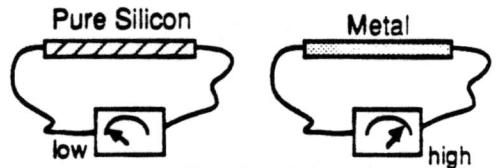

Conductivity

CONDUCTOR: a material which has high conductivity (i.e. low resistivity). Common conductors are found in group IV of the periodic table.

Conductors

22

CONTACT: a region of device where conductive material meets the semiconducting material substrate enabling electrical access to the IC components through the chip package.

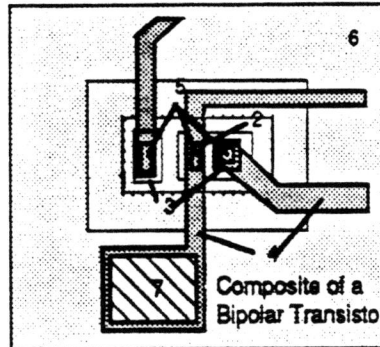

1. Buried Layer
2. Base
3. Emitter
4. Metal
5. Contact
6. Passivation
7. Bonding Pad

Composite of a Bipolar Transistor

CONTACT ALIGNER: an image transfer system in which the wafer is in physical contact with the mask.

Wafer — Mask
Resist
Chuck

Contact Alignment

CONTACT MASK: (1) the step at which holes are put into the wafer layers to allow the metal layer to reach down to the doped silicon substrate, (2) photoplate used to transfer an image to a resist coated wafer optically, mask and wafer touch.

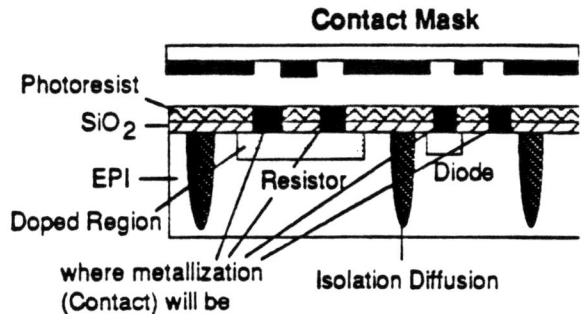

Contact Mask

Photoresist
SiO_2
EPI
Doped Region

Resistor
Diode

where metallization (Contact) will be
Isolation Diffusion

CONTAMINATION: a general term used to describe unwanted material that adversely affects the physical or electrical characteristics of a semiconductor wafer.

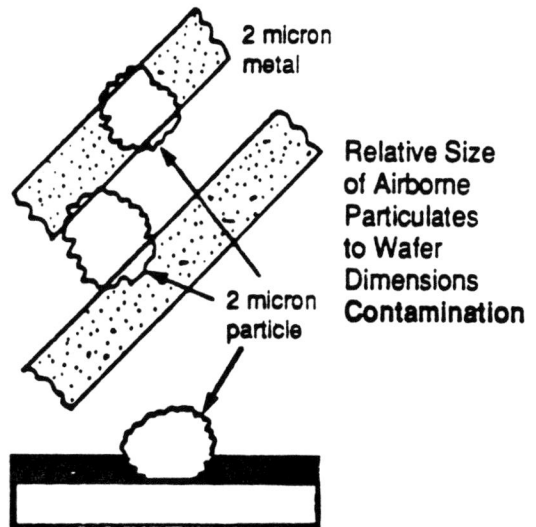

2 micron metal

2 micron particle

Relative Size of Airborne Particulates to Wafer Dimensions **Contamination**

CONVECTION: one of the three methods of heat transfer (conduction, convection and radiation). This type of heating is accomplished through transfer of heated gas. The gas is heated in one location and transferred to the object to be heated. Examples of convection heating are home forced-air heating and hair dryers.

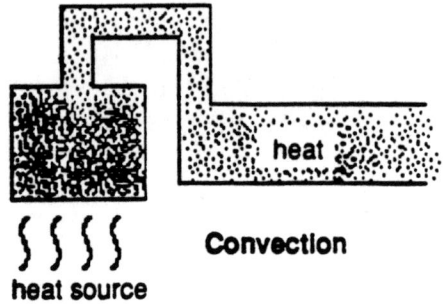

heat

heat source

Convection

CRITICAL DIMENSIONS: the widths of the lines and/or spaces of circuit patterns as well as the area of contact; usually refers to the smallest features in the pattern.

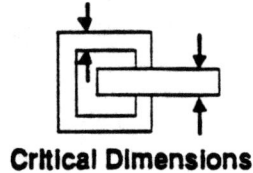

Critical Dimensions

To vacuum system

Stage 1
Cold Surface

CYROGENIC PUMP: an evacuation pump that can produce a vacuum to the 10^{-10} Torr range, the same level as the vacuum of space. This design doesn't require forepumps or cold traps and is faster than other types of vacuum pumps.

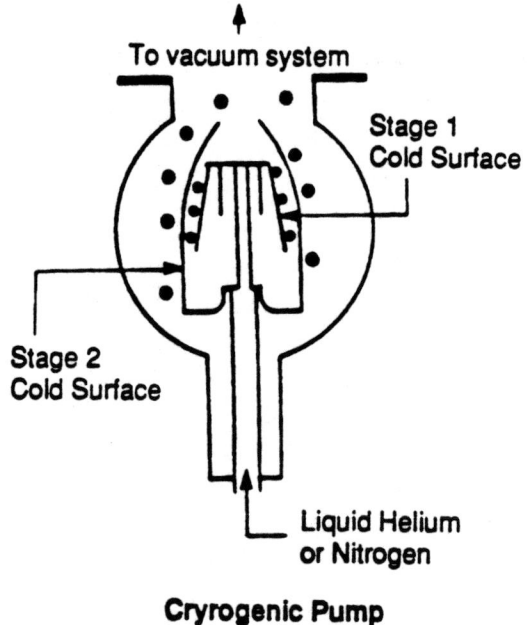

Stage 2
Cold Surface

Liquid Helium
or Nitrogen

Cryogenic Pump

24

CRYSTAL: a material in which the atoms are arranged in structured groups called unit cells.

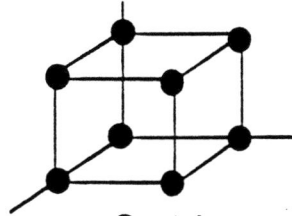
Crystal

CRYSTAL DEFECTS: defects such as vacancies and dislocations in a crystal which influence the electrical performance of a circuit.

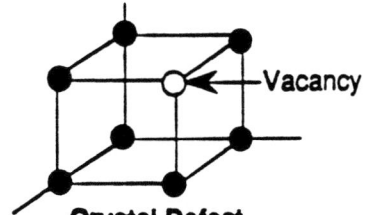
←Vacancy
Crystal Defect

CRYSTAL GROWTH DEFECTS: conditions occurring during the growth of semiconductor grade silicon which result in gross structural defects. Two common types are slip and twining.

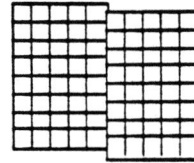
Slip
Crystal Growth Defect

CRYSTAL PLANES: the planes in the semiconductor crystal structure along which the die must be aligned in order to prevent "ragged" die edges when the wafer is separated into individual die.

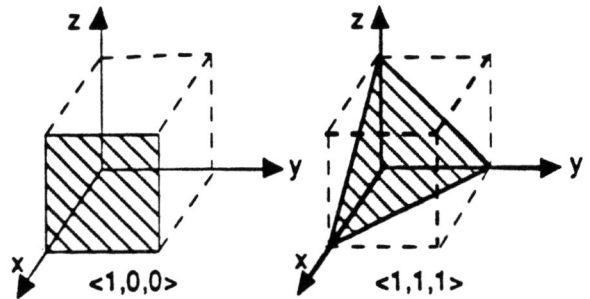
<1,0,0>   <1,1,1>
Crystal Planes
Miller Indices <1,0,0> and <1,1,1>

CUM YIELD: see "Fabrication Yield"

CURRENT: a measure of the rate of flow of electrical charge. Measured in units of "amps".

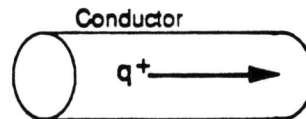
Conductor
q$^+$ ——→

i = Current = q/t
q = amount of positive charge
t = time

CURVE TRACER: electrical test equipment that displays the characteristics of a device visually on a screen.

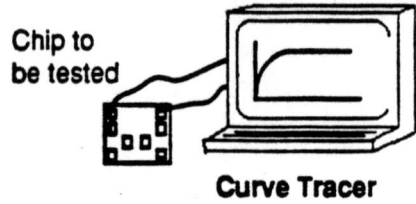

Chip to be tested

Curve Tracer

CZOCHRALSKI (CZ) METHOD: a crystal growth method for producing semiconductor grade silicon from polycrystalline silicon. Pieces of polysilicon and dopant are heated in a crucible. A seed crystal is touched to the molten surface and slowly raised resulting in the growth of a cylindrical ingot having the same crystal orientation as the seed

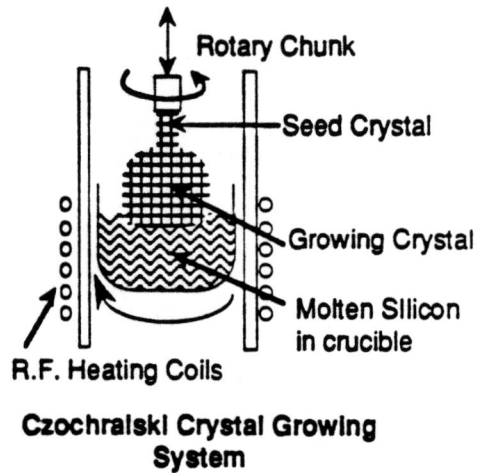

Rotary Chunk

Seed Crystal

Growing Crystal

Molten Silicon in crucible

R.F. Heating Coils

Czochralski Crystal Growing System

26

DI: see "Develop Inspection".

DI Water: see "Deionized Water".

DIP: see"Dual In-Line Package".

DRAM: see "Dynamic RAM".

DUV: see "Deep Ultraviolet".

DARK FIELD MASK: a mask on which the pattern is defined by the clear portion of the mask.

Dark Field Mask

DEEP ULTRAVIOLET: A light wavelength often used to expose photoresist which has the advantage of an ability to produce smaller image widths. (wavelength approximately 225 nm)

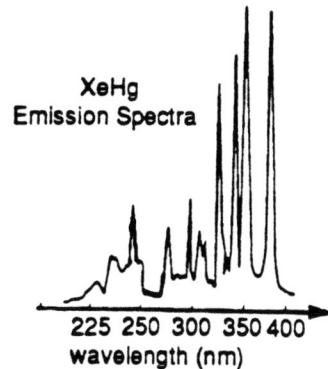

XeHg
Emission Spectra

225 250 300 350 400
wavelength (nm)

Deep UV

DEGLAZE: the process in which the thin surface layer of doped silicon dioxide is removed from wafers coming out of deposition to remove unwanted oxide and dopant sources. It usually consists of an HF dip.

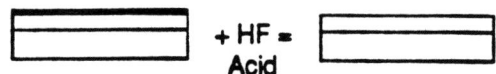

+ HF =
Acid

Deglaze

DEHYDRATION BAKING: a heating process by which wafer surfaces are restored to a hydrophobic condition by baking. Surface water is evaporated from the wafer at elevated temperatures.

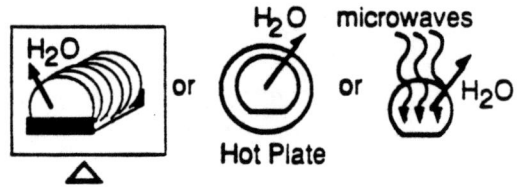

Dehydration Baking

DEIONIZATION: a process by which ions are removed from water to make it non-conducting.

DEIONIZED WATER: the purity of this water is measured by its resistivity with the standard being 18 MOhm

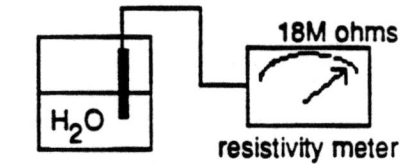

Deionized Water

DEMINERALIZATION: see "Deionization"

DEPLETION REGION: the region in a semiconductor where essentially all charge carriers have been swept out by the electric field which exists there.

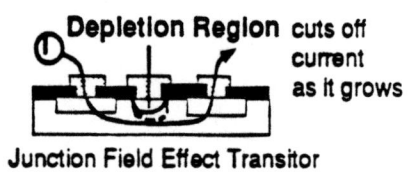

Depletion Region cuts off current as it grows

Junction Field Effect Transitor

DEPOSITION: process in which layers are formed as the result of a chemical reaction in which the desired layer material is formed and coats the wafer surface.

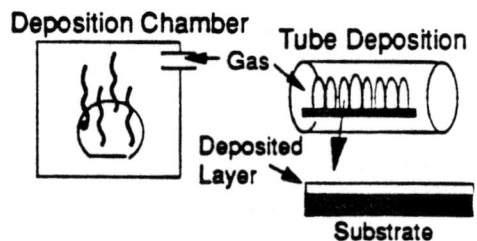

Deposition

DEVELOP INSPECTION: the first inspection in the wafer patterning process. It consists of measurement of critical dimensions and inspection for defects. It is done after development or development/hard bake if an automatic baking system is used.

Develop Inspection

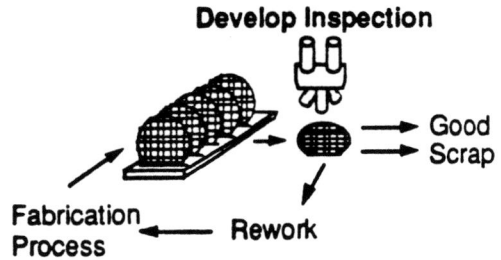

DEVELOPER: a chemical used to remove areas defined in the masking and exposure step of wafer fabrication.

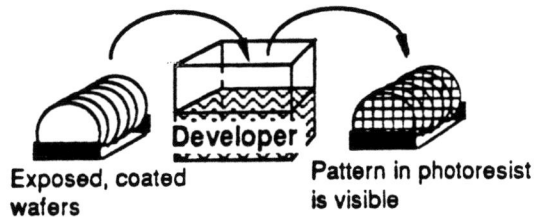

Exposed, coated wafers

Pattern in photoresist is visible

DEVELOPMENT: the processing step in which photoresist is removed from areas defined by the masking and exposure step of wafer fabrication.

Before Develop

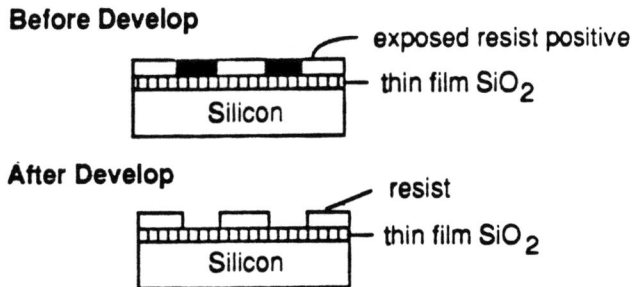

exposed resist positive

thin film SiO_2

Silicon

After Develop

resist

thin film SiO_2

Silicon

DEVICE: a discrete circuit (such as a transistor, resistor, or capacitor) or an integrated circuit .

DIBORANE (B_2H_6): a gas that is often used as a source of boron for doping silicon.

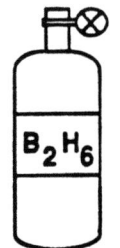

B_2H_6

Diborane

DICHLOROSILANE: a primer chemical which bonds to both the wafer surface and photoresist layer.

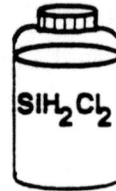

Dichlorosilane

DIE: one unit on a wafer separated by scribe lines. After all of the wafer fabrication steps are completed, die are separated by sawing, the separated units are referred to as "chips"

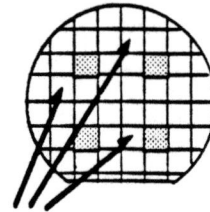

Die

DIE BONDING: an assembly step in which individual chips are attached to the the package with conductive adhesives or metal alloys.

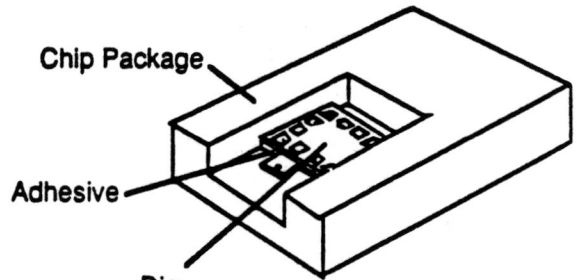

Chip Package

Adhesive

Die

Die Bonding

DIE FIT: the layer-to-layer alignment of die patterns

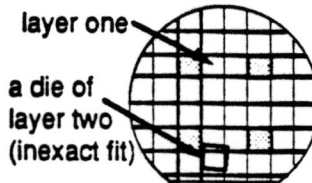

DIE SORT: see "Wafer Sort".

layer one

a die of layer two (inexact fit)

Die Fit

DIELECTRIC: a material that conducts no current when a voltage is applied. Two common dielectrics encountered in semiconductor processing are silicon dioxide and silicon nitride.

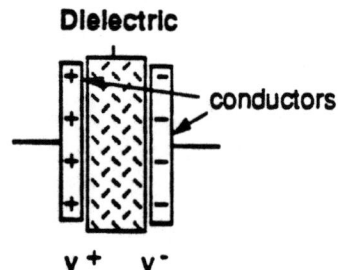

Dielectric

conductors

v+   v−

30

DIFFRACTION: the bending of light around the mask pattern. It is a major limiting factor in image resolution.

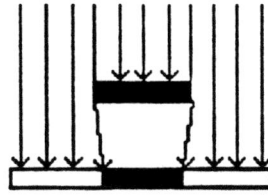
Diffraction

DIFFUSION: a process used in wafer fabrication which introduces minute amounts of dopants (impurities) into a substrate material such as silicon or germanium and permits the impurity to spread into the substrate. The process is dependent on temperature and time.

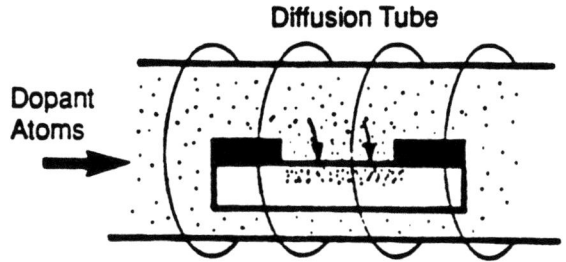
Diffusion Tube

Dopant Atoms

Resistance Coils

Solid State Diffusion

DIFFUSIVITY: the rate of movement or diffusion of dopants in semiconducting materials.

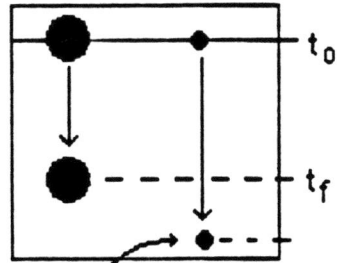
t_0

t_f

Particle with greater **Diffusivity**

DIGITAL: a circuit which uses two states (on/off) to convey information. A digital display gives information in measured amounts.

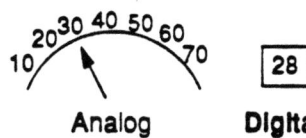
Analog   **Digital**

Data Representation Schemes

DIGITIZING: procedure in which a drawing is traced by a cursor, reduced to a bit map, and stored digitally in memory. This information is used to produce reticles or drive E-beam systems.

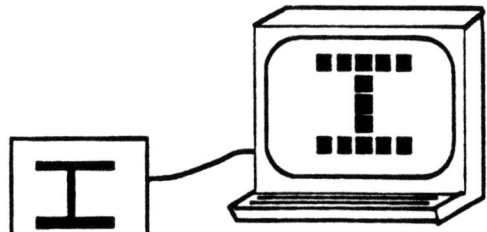
Digitizing

31

DIODE: device which enables current flow in one direction but not in the other.

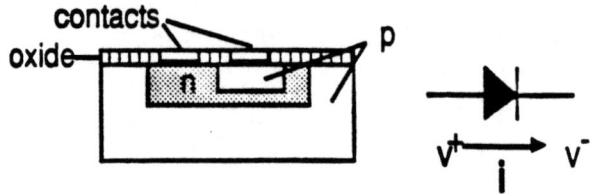

contacts
oxide
p
n

V^+ → i → V^-

Diode

DIRECT CURRENT: electrons flowing in one direction (DC).

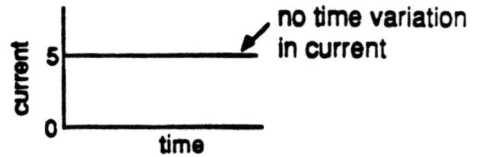

no time variation in current

current
5
0
time

Direct Current

DISCRETE DEVICE: a circuit having a single electrical function. Discrete devices include capacitors, resistors, transistors, diodes and fuses.

Discrete Device

DISLOCATION: a type of crystal defect resulting from a discontinuity in the crystal lattice.

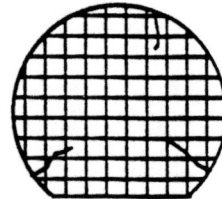

Dislocations

DMOS: Diffused MOS. A transistor structure which features a narrow (channel length) separation between the source and drain. This structure is created by two sequential diffusions through the same hole.

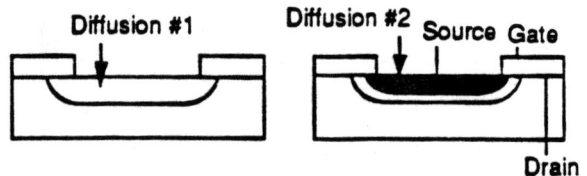

Diffusion #1   Diffusion #2   Source   Gate

Drain

DMOS

32

DONOR: a dopant (impurity) that can make semiconducting materials N-type by providing extra "free" electrons. A donor replaces a Si atom in the crystal structure. Because a donor has one more electron than the silicon, this elecron is free to move in the crystal.

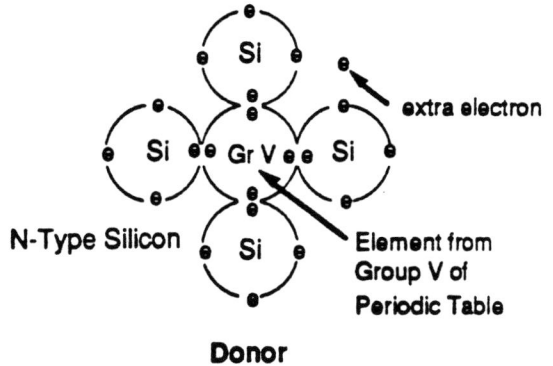

Donor

DOUBLE MASKING: a process in which photoresist coating is applied twice for each patterned layer enabling the use of thinner films. Used to eliminate pinholes in highly integrated circuits.

1. 1st layer processed to develop

2. 2nd photoresist layer applied and exposed with oversize mask

3. 2nd layer developed

4. After etch and strip

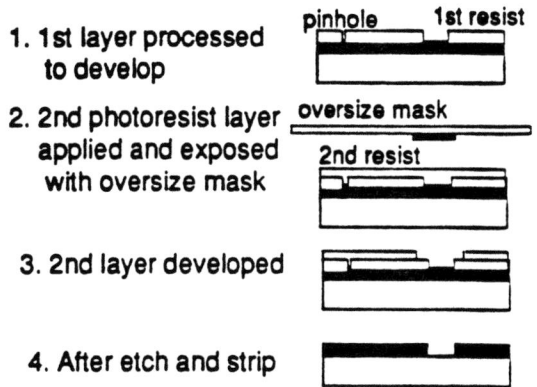

Double Masking

DOPANT: an element that alters the conductivity of a semiconducting material by contributing a free hole (P-type) or electron (N-type) to the crystal. Dopants are found in Groups III and V of the Periodic Table.

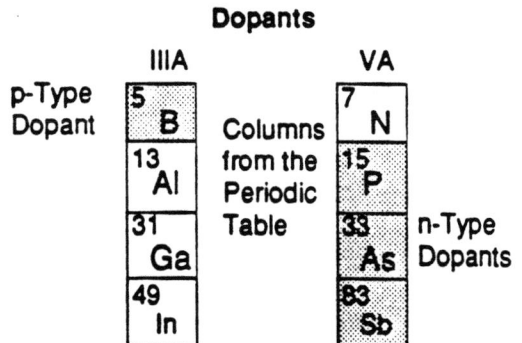

Dopants

DOPING: the introduction of a dopant (impurity) by diffusion or ion implantation into the crystal structure of semi-conducting materials. The doping process can be used to tightly control conductivity because conductivity in the crystal is directly proportional to the amount of dopant. An example of doping is adding boron to silicon to make the material P-type.

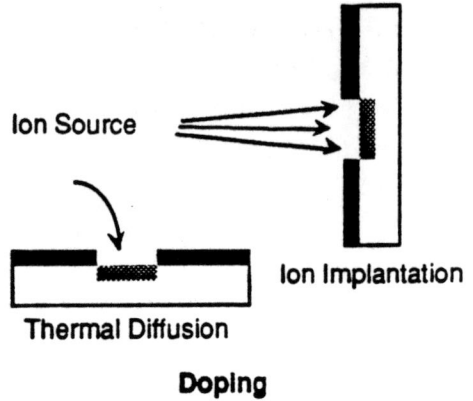

Thermal Diffusion

Ion Implantation

Doping

DOPING SEQUENCE: a three-step sequence which results in the formation of the localize N or P-type regions in the wafer surface. The basic sequence of operations is: Layering, Patterning, Doping.

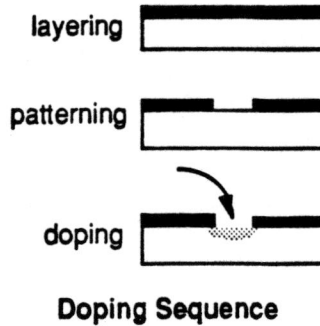

layering

patterning

doping

Doping Sequence

DRAIN: one of the three regions of a unipolar or field effect transistor (FET), along with the source and gate.

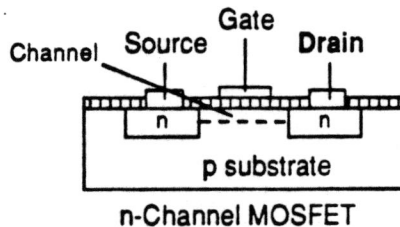

Channel Source Gate **Drain**

n n

p substrate

n-Channel MOSFET

DRIVE-IN: the stage in diffusion where the dopant is positioned deeper into the

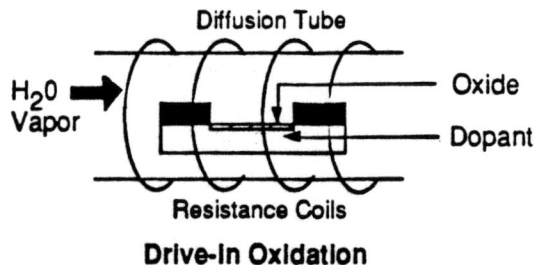

Diffusion Tube

H_2O Vapor

Oxide

Dopant

Resistance Coils

Drive-In Oxidation

DRY ETCHING: a process resulting in the selective removal of material achieved by use of gas.

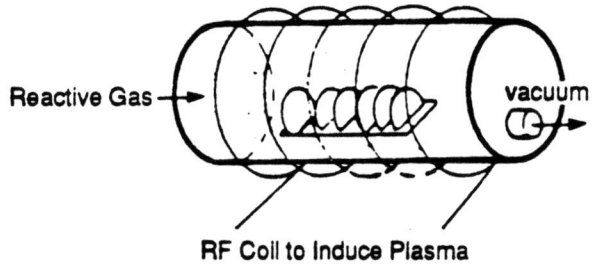

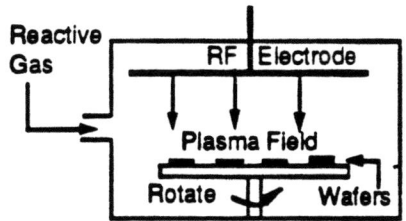

Reactive Gas →

vacuum

RF Coil to Induce Plasma

Reactive Gas

RF Electrode

Plasma Field

Rotate → Wafers

Barrel and Planar Plasma (Dry) Etching

DRY OX: the growth of silicon dioxide using oxygen and hydrogen (which form water vapor at process temperatures) rather than using water vapor directly.

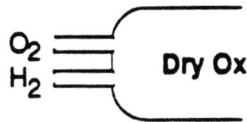

O_2
H_2

Dry Ox

DRY OXIDE: silicon dioxide grown using oxygen.

DRY STEAM: see "Dry Ox".

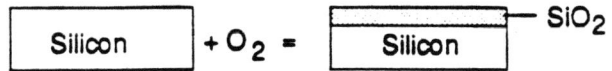

Silicon $+ O_2 =$ Silicon — SiO_2

Dry Oxide

DUAL IN-LINE PACKAGE (DIP): a rectangular integrated circuit package, with leads coming out of the long sides and bent down to fit into a socket.

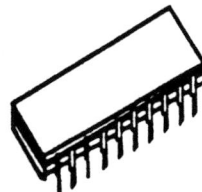

Dual In-Line Package (DIP)

DUAL LAYER PHOTORESIST:

successive applications of thin layers of photoresist .

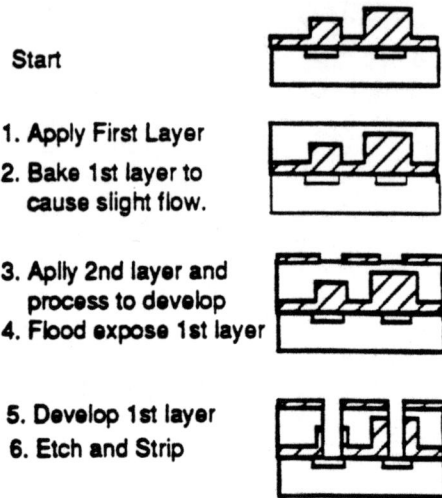

Start

1. Apply First Layer
2. Bake 1st layer to cause slight flow.

3. Aplly 2nd layer and process to develop
4. Flood expose 1st layer

5. Develop 1st layer
6. Etch and Strip

Dual Layer Photoresist Processing

DYNAMIC RAM: memory

device in which the charge must be refreshed to maintain information.

E

ECL: Emitter Coupled Logic device.

EEPROM: see "Electrically Erasable Programmable Random Access Memory".

EPROM: see "Erasable Programmable Read Only Memory".

E-BEAM: an exposure source which allows direct image formation without a mask. An E-beam can be deflected by electrostatic plates and directed to precise locations resulting in the generation of sub-micron patterns.

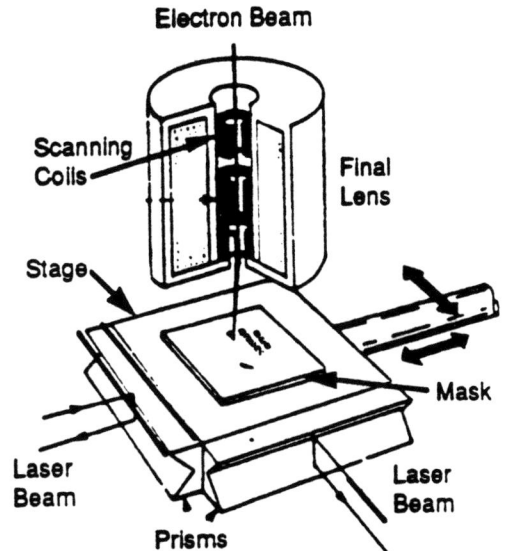

Electron Beam
Direct Write Apparatus
for Mask Making

E-BEAM EVAPORATION: (Electron Beam Evaporation) phase change that uses the energy of a focused electron beam to provide the required energy to change solid metal or alloys from solid to gas.

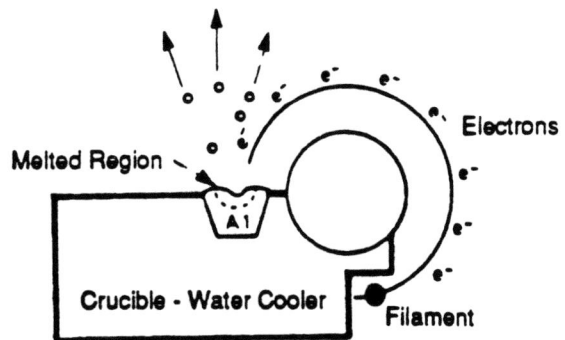

E-Beam Evaporation

37

**E-BEAM EXPOSURE SYS-
TEM**: a machine in which
the image pattern is stored
in a computer memory and
used to control the electrostatic plates that in turn
direct a beam of electrons
resulting in the generation of
patterns without the use of
reticles or photomasks.

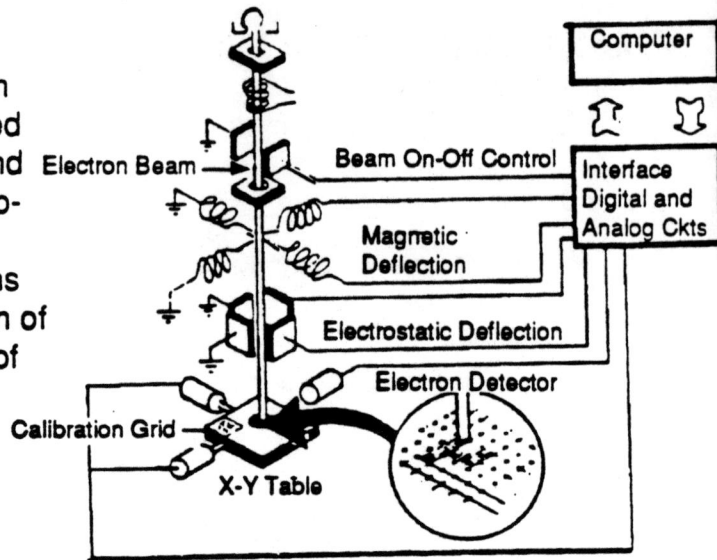

E-TEST: see "Wafer Sort".

EDGE DIE: the incomplete die
located on the edge of the
wafer

ELECTRICAL TEST: see
"Wafer Sort".

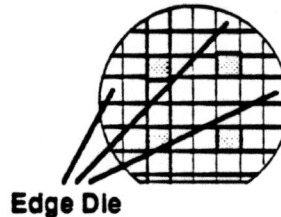

**ELECTRICALLY ERASABLE
PROGRAMMABLE RAN-
DOM ACCESS MEMORY
(EEPROM)**: a device,
similar to ROM with capability of selective erasure of
memory. Sometimes re-
ferred to as E²ROM or
EEROM.

**ELECTROMAGNETIC SPEC-
TRUM**: the full range of
radiation wavelengths that
vary from very short (x-rays)
to very long (radio wave)
and includes infra-red,
visible and ultraviolet light.

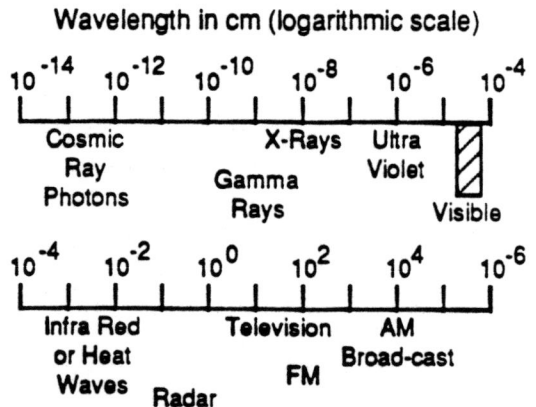

Computer

Electron Beam — Beam On-Off Control — Interface Digital and Analog Ckts

Magnetic Deflection

Electrostatic Deflection

Electron Detector

Calibration Grid

X-Y Table

E-Beam Exposure System

Edge Die

Wavelength in cm (logarithmic scale)

10^{-14} 10^{-12} 10^{-10} 10^{-8} 10^{-6} 10^{-4}

Cosmic Ray Photons   X-Rays   Ultra Violet

Gamma Rays

Visible

10^{-4} 10^{-2} 10^{0} 10^{2} 10^{4} 10^{-6}

Infra Red or Heat Waves   Television   AM Broad-cast

Radar   FM

Electromagnetic Spectrum

ELECTROMIGRATION: the diffusion of electrons in electric fields set up in the thin film metal wiring while the circuit is in operation. It occurs in aluminum and is exhibited as a field failure, not as a process defect. The metal thins and eventually separates completely, causing an opening in the circuit.

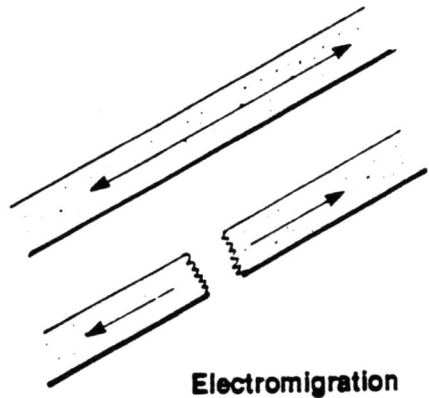

Electromigration

ELECTRON: a negatively charged particle revolving around the nucleus of an atom. If an atom gains or loses an electron, it becomes an ion.

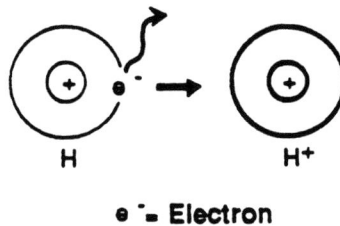

H   H+

e ⁻ = Electron

ELECTRON BEAM: see "E-Beam".

ELECTRONIC GRADE SILICON: purified silicon in which the impurity level has been reduced to a level of no more than several parts per million.

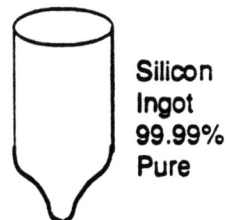

Silicon
Ingot
99.99%
Pure

Electronic Grade Silicon

ELEPHANT: a fixture which mates to the quartz oxidation tube into which a boat of wafers is placed.

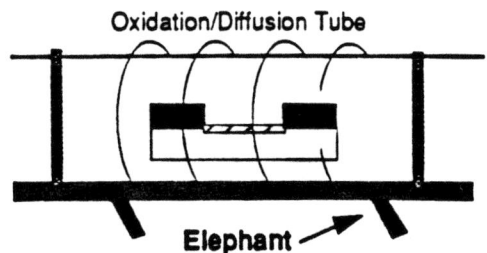

Oxidation/Diffusion Tube

Elephant

ELLIPSEOMETER: instrument that uses laser light sources to measure thin film thickness.

EMITTER: 1. The region of a transistor that serves as the source or input end for carriers. 2. The N-type diffusion usually done using phosphorus which forms the emitter of NPN transistors, the base contact of PNP transistors, the N^+ contact of NPN transistors and low value resistors.

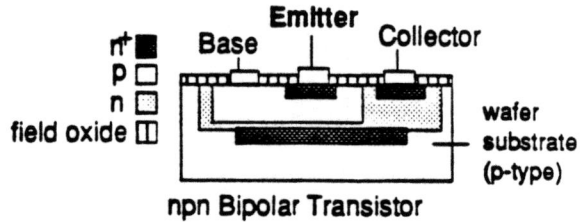

npn Bipolar Transistor

EMULSION: type of photomask in which the dark area is formed from a suspension of an opaque salt .

ENCAPSULATION: the process of enclosing a chip in its package.

ENGINEERING TEST DIE: see "Test Die".

ENIAC (Electronic Numerical Integrator and Calculator): the world's first electronic computer.

Eniac Statistics:
weight: 50 tons
size: 300 sq. ft.
vacuum tubes: 19,000
needed its own small power station

EPI: see "Epitaxial".

40

EPITAXIAL (Greek for "arranged upon"): the growth of a single crystal semiconductor film upon a single crystal substrate. The epitaxial layer has the same crystallographic characteristics as the substrate material.

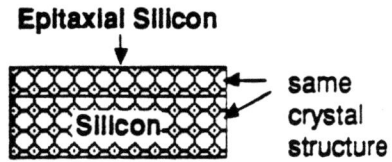

Epitaxial Silicon

same crystal structure

Silicon

EPOXY: group of resins used to adhere chips to their packages.

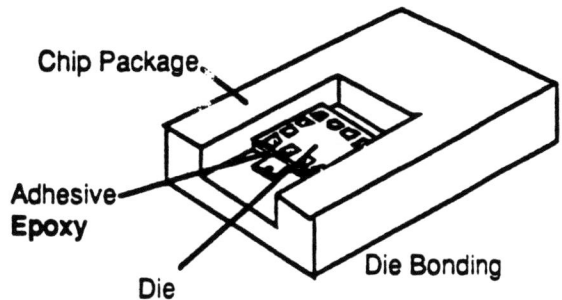

Chip Package

Adhesive Epoxy

Die

Die Bonding

ERASABLE PROGRAMMABLE READ ONLY MEMORY (EPROM): a device that allows stored information to be erased. Erasing is typically accomplished with ultraviolet light .

ETCH: a process for removing material in a specified area through a chemical reaction.

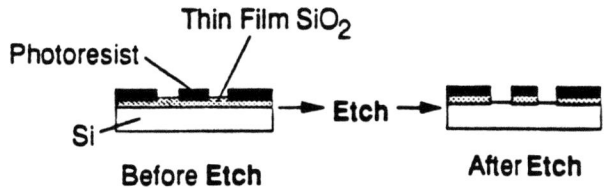

Photoresist

Thin Film SiO_2

Si

→ Etch →

Before Etch

After Etch

EUTECTIC POINT: temperature at which a metal and semiconducting material that are in contact with each other melt. This temperature is lower than the melting temperature of the individual materials.

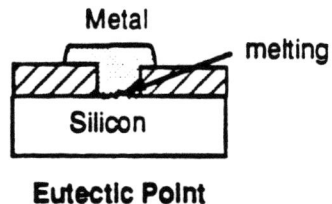

Metal

melting

Silicon

Eutectic Point

EVAPORATION: a process step that uses heat to change a material (usually a metal or metal alloy) from its solid state to gaseous state resulting in the deposition of a thin layer of metal or metal alloy on the wafer surface. Both electron beam and filament evaporation are common in semiconductor processing.

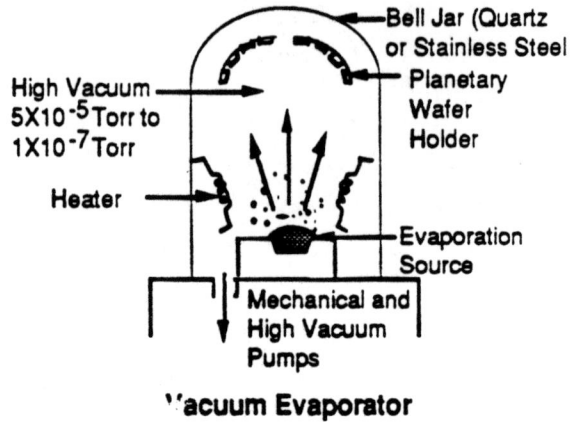

Bell Jar (Quartz or Stainless Steel

High Vacuum 5×10^{-5} Torr to 1×10^{-7} Torr

Planetary Wafer Holder

Heater

Evaporation Source

Mechanical and High Vacuum Pumps

Vacuum Evaporator

EXPOSURE: method of defining patterns by the interaction of light (or other form of energy) with light sensitive photoresist.

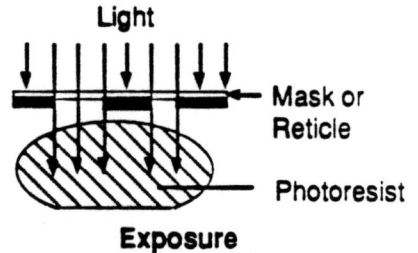

Light

Mask or Reticle

Photoresist

Exposure

EXPOSURE SPEED: the speed with which photoresist reacts to exposure from an energy source.

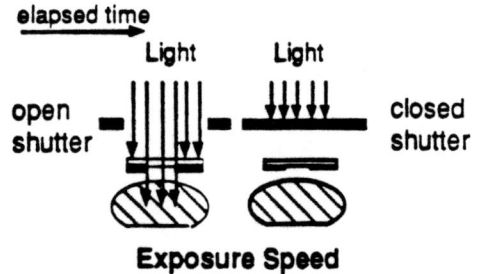

elapsed time

Light   Light

open shutter   closed shutter

Exposure Speed

EXTRUSION: see "Pattern Defects"

F

FAB: see " Wafer Fab".

FAB YIELD: see "Fabrication Yield".

FABRICATION: integrated circuit manufacturing consisting of layering, patterning, doping, and heat treating.

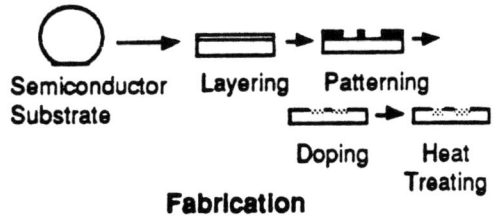

Fabrication

FABRICATION YIELD: the percent of wafers arriving at wafer sort compared with the number started into the process.

$$\frac{\text{\# Good Wafers to Wafer Sort}}{\text{\# Wafers Started}} \times 100\% =$$

Fabrication Yield

FAILURE ANALYSIS: procedure for determining the cause(s) of device failure

FAILURE RATE: a measure of reliability expressed as failure per unit of time, typically recorded in failures per 1000 hrs.

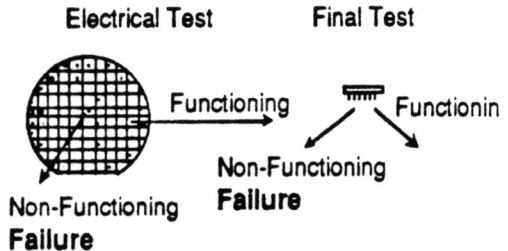

Failure Analysis

Failure Rate = # of Failures / 1000 hours

FET (Field-Effect Transistor): a transistor consisting of a source, gate and drain, whose action depends on the flow of majority carriers past the gate from the source to drain. The flow is controlled by the transverse electric field under the gate. (See "Unipolar Transistor").

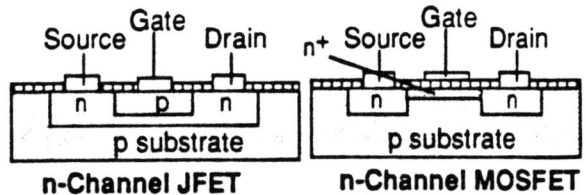

n-Channel JFET **n-Channel MOSFET**

43

FIELD OXIDE: the region on a electrical device where the oxide serves the function of a dielectric.

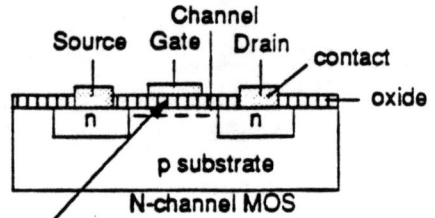

Channel
Source Gate Drain — contact
— oxide
n   n
p substrate
N-channel MOS

Field Oxide

FILAMENT: a coiled piece of wire that is in contact with a material to be evaporated and is heated by passing current through it.

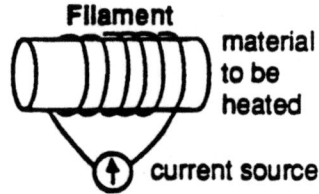

Filament
material to be heated
current source

FILAR MEASURING EYE-PIECE: attachment to microscope used for measuring critical dimensions

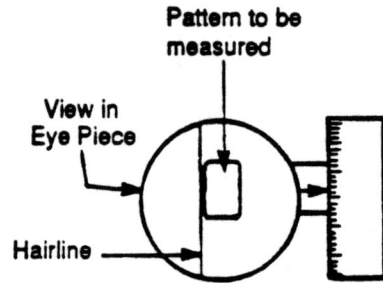

Pattern to be measured
View in Eye Piece
Hairline

Manual Filar Measurement

FILTERED NITROGEN BLOW-OFF: method of particulate removal in which surfaces are flushed with filtered nitrogen.

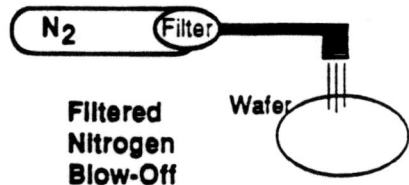

N_2 Filter
Wafer
Filtered Nitrogen Blow-Off

FINAL INSPECTION: last inspection after wafer fabrication process is completed. Used to determine if the wafer will proceed into the assembly area.

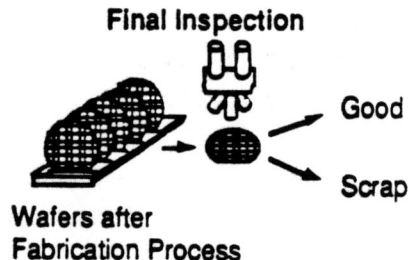

Final Inspection
Good
Scrap
Wafers after Fabrication Process

44

FINAL TEST: the final assembly step in which the packaged chip is evaluated for electrical functionality.

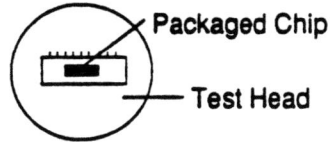
Packaged Chip
Test Head

FLASH POINT: the temperature above which solvent vapors will ignite in the presence of an open flame.

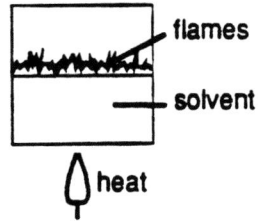
flames
solvent
heat
Flash Point

FLASH SYSTEM: a system in which small portions of water are dropped on a hot plate and immediately converted to vapor and carried into an oxidation tube by a carrier gas.

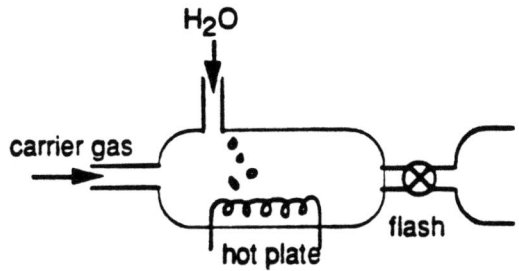
H_2O
carrier gas
flash
hot plate
Flash System

FLAT PACK: a ceramic, surface-mounted, hermetically sealed chip package.

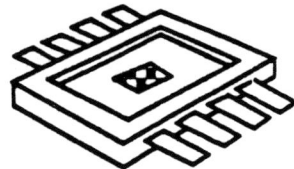
Flat Pack

FLAT ZONE: the region of an oxidation furnace which is highly temperature controlled.

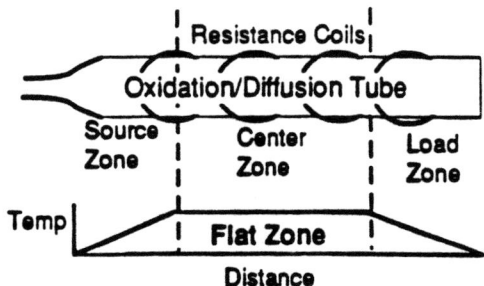
Resistance Coils
Oxidation/Diffusion Tube
Source Zone
Center Zone
Load Zone
Temp
Flat Zone
Distance

FLOAT ZONE TECHNIQUE: a polycrystalline silicon processing technique used to produce silicon containing less oxygen impurity than can be produced using the CZ method. In a chamber of inert gas, a bar of polycrystal material with a seed attached to its end is moved past by an RF coil which melts the atoms of the polycrystalline material and then causes them to solidify with the same orientation as the seed .

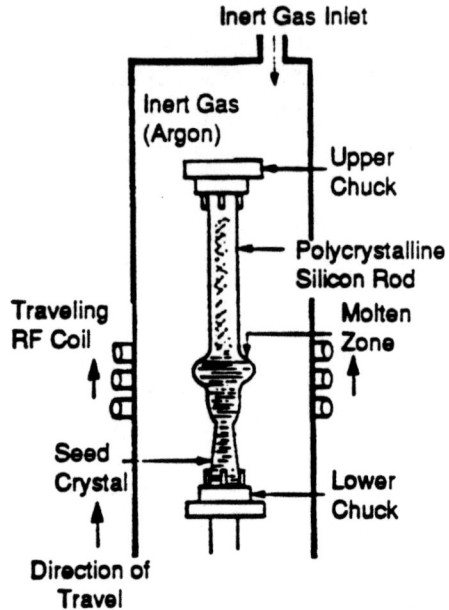

Float Zone Crystal Growing System

FOUR-POINT PROBE: a piece of electrical test equipment used to determine the sheet resistivity of a wafer.

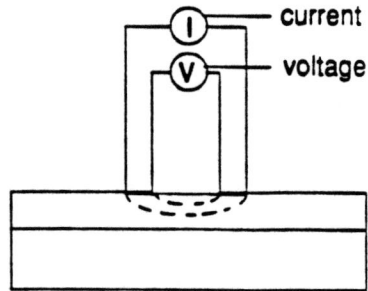

4-Point Probe measurement of a thin layer

FREQUENCY: number of wave cycles per unit of time. Used to measure number of times alternating current completes a cycle. Unit is Hertz(Hz) which is equal to one cycle per second.

Frequency = number waves/unit time

46

FRINGES: approximately parallel color bands used to visually determine film thickness.

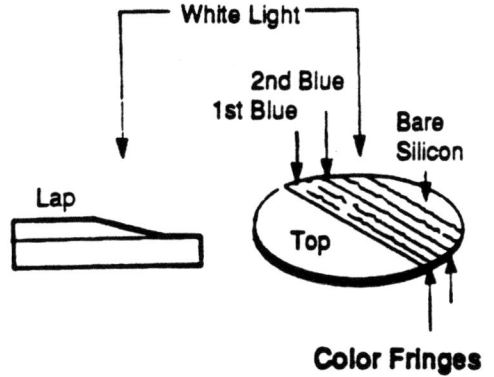

White Light

2nd Blue
1st Blue

Bare Silicon

Lap

Top

Color Fringes

FURNACE: a piece of equipment containing a resistance heated element and a temperature controller. It is used to maintain a region of constant temperature with a controlled atmosphere for the processing of semiconductor devices.

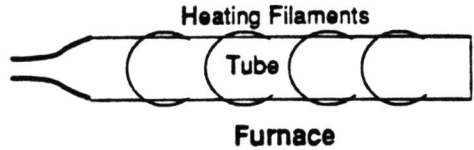

Heating Filaments

Tube

Furnace

FUSE: a circuit component that functions as a protection device. In semi-custom products, unnecessary circuits are removed by causing fuses to malfunction ("blow").

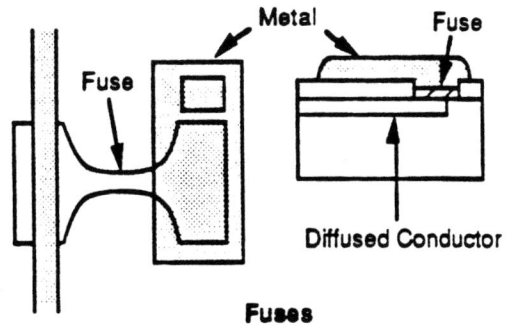

Metal

Fuse

Fuse

Fuse

Diffused Conductor

Fuses

G

GALLIUM ARSENIDE (GaAs):
Most common of compound
semiconductor materials. It
has the advantage of pro-
ducing higher speed devices
than those made using
silicon as a substrate.

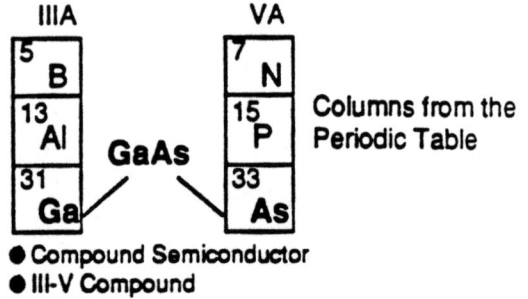

IIIA   VA

5 B		7 N
13 Al	GaAs	15 P
31 Ga		33 As

Columns from the
Periodic Table

● Compound Semiconductor
● III-V Compound

GATE: one of the three regions
of the unipolar or field-effect
transistor (FET or MOS),
along with the source and
drain.

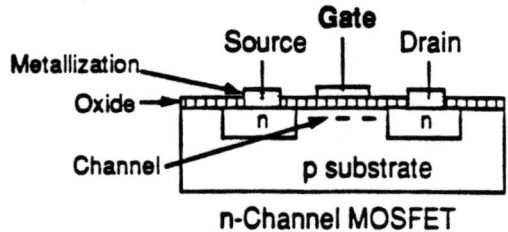

Gate

Metallization   Source   Drain

Oxide →

Channel →   n   n   p substrate

n-Channel MOSFET

GATE ARRAY: a type of
integrated circuit made up of
an arrangement of interconnected gates used to pro-
vide custom functions.

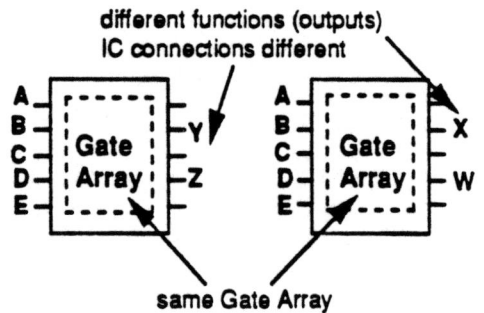

different functions (outputs)
IC connections different

A
B
C   Gate
D   Array
E   Y
Z

A
B   Gate
C   Array
D   X
E   W

same Gate Array

GATE OXIDE (Gate Ox): the
thin oxide which causes the
induction of charge, creating
a channel between source
and drain regions of a MOS
transistor.

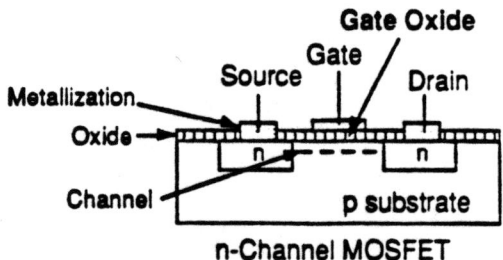

Gate Oxide

Source   Gate   Drain

Metallization

Oxide →

Channel →   n   n   p substrate

n-Channel MOSFET

48

GAUSSIAN DISTRIBUTION: a typical distribution of random data about a median value.

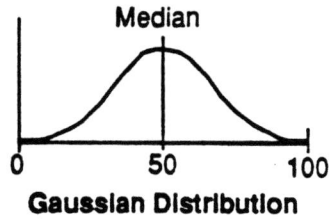
Gaussian Distribution

GERMANIUM: semiconducting material used in the manufacture of crystal diodes and of early transistors

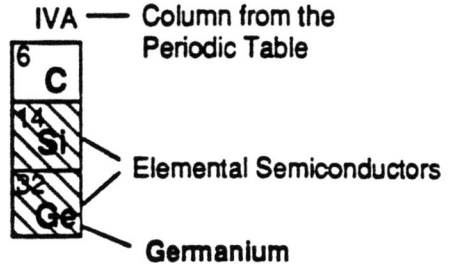

GIGA: prefix used to denote one billion.

$$Giga = 10^9$$

GLASSIVATION: alternate name for passivation layer consisting of silicon dioxide (glass) used to protect and seal the wafer surface.

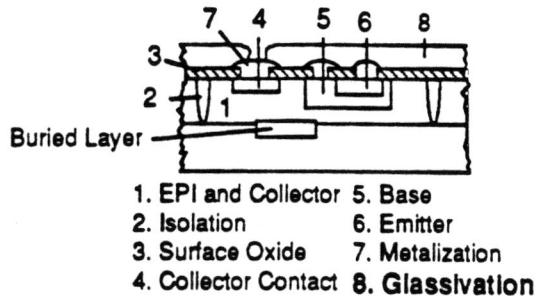

1. EPI and Collector   5. Base
2. Isolation   6. Emitter
3. Surface Oxide   7. Metalization
4. Collector Contact   8. **Glassivation**

GROWN JUNCTION: P/N junctions made by controlling the type of impurity in a single crystal while it is being grown from a melt.

GROWTH DEFECTS: see "Crystal Growth Defects".

H

HEPA FILTER: see "High Efficiency Particulate Attenuator".

HF: 1) see "High Frequency"
2) see "Hydrofluoric Acid".

HMDS: see "Hexamethyldisilizane".

HMOS: High Density MOS.

HARDWARE: the physical components of a computer or electronic system, differentiated from system software.

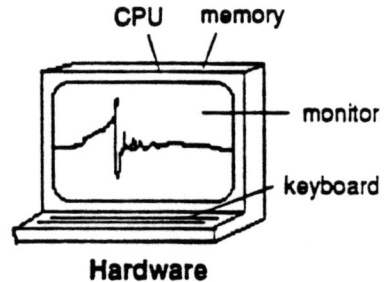

Hardware

HEXAMETHYLDISILIZANE (HMDS): a primer used to promote photoresist adhesion.

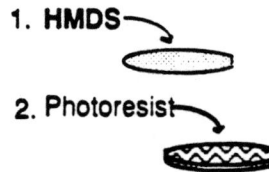

HI REL: (High Reliability) used to refer to circuits which receive stringent environmental testing for conditions such as high temperature and humidity and mechanical shock.

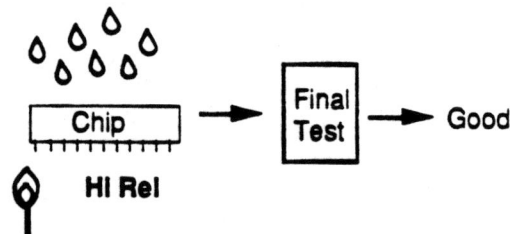

50

**HIGH EFFICIENCY PARTICU-
LATE ATTENUATOR (HEPA)
FILTER:** a filter constructed of
fragile fibers in an accordion
folded design which allows a
large filtering area at an air ve-
locity low enough for operator
comfort. This filters permit a
filtering efficiency of 99.99%.

HEPA Filter Design

HIGH FREQUENCY (HF): the
type of heating created by
bombardment with high frequency waves. HF heating is
used in many silicon operations
to obtain high temperatures.

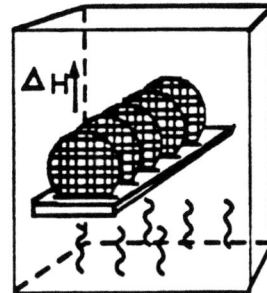

**High Frequncy (HF)
Heating**

HIGH PRESSURE OXIDATION:
oxidation carried out at high
pressure (10-20 atms) is used
to reduce the amount of heat
required. The reaction chamber
for this process must be constructed of stainless steel to
safely contain the pressure.

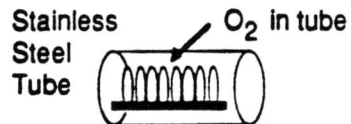

High Pressure Oxidation

**HIGH PRESSURE WATER
CLEANING:** a high pressure
stream of water (2000 = 4000
lbs/in²) is swept across the
wafer or mask to dislodge
particulates. Drying is accomplished by high-speed rotation.

**High
Pressure
Water
Cleaning**

HOLE: 1) the absence of a valence electron in a semiconductor crystal (see acceptor). Motion of a hole is equivalent to motion of a positive charge. 2) a region in which material is missing.

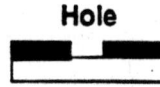

Hole

HYBRID INTEGRATED CIRCUIT: a structure consisting of an assembly of one or more semiconductor devices and a thin-film integrated circuit on a single substrate, usually of ceramic.

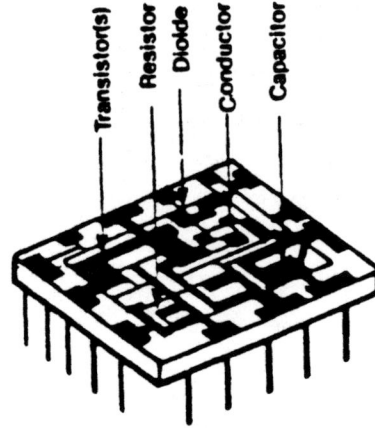

Hybrid Circuit

HYDROFLUORIC ACID (HF): an acid used to etch silicon dioxide which is often diluted or buffered before use.

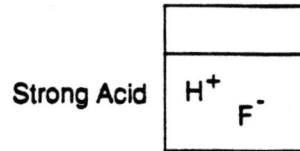

Strong Acid H^+ F^-

example: HF

HYDROGEN (H_2): a gas used in semiconductor processing primarily as a carrier gas for high temperature reaction steps such as epitaxial silicon growth.

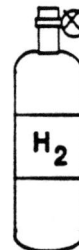

Hydrogen

HYDROPHILIC: affinity toward water (water loving). A hydrophilic surface is one that will allow water to spread across it in large puddles.

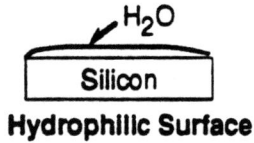

Hydrophilic Surface

HYDROPHOBIC: aversion to water. A hydrophobic surface is one that will not support large pools of water on its surface. The water is pulled into droplets on the surface. These surfaces often are termed "dewetted".

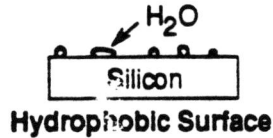

Hydrophobic Surface

HYDROSCOPIC: attracts and absorbs water. A hydroscopic surface pulls water into itself.

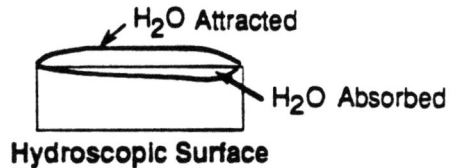

Hydroscopic Surface

I

IC: see "Integrated Circuit"

IMAGE: 1) to transfer a pattern onto wafer surface using optical methods 2) the pattern on the wafer's surface is sometimes referred to as the image.

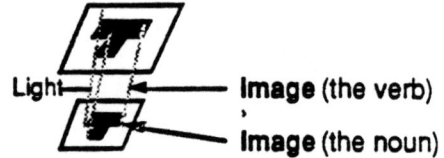

Light ── **Image** (the verb)

── **Image** (the noun)

IMPURITY: refers to small amounts of dopant materials introduced into a semiconductor substrate.

INDEX OF REFRACTION: a number which indicates the amount that light will bend as it enters a particular material.

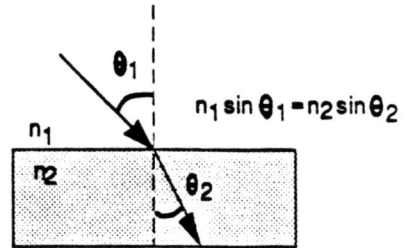

θ_1

$n_1 \sin\theta_1 = n_2 \sin\theta_2$

n_1

n_2

θ_2

n = Index of Refraction

INGOT: material prepared by solidification from a melt, wafers are made from ingots.

Silicon
Ingot
99.99%
Pure

INSULATED GATE FIELD EFFECT TRANSISTOR: see "MOSFET".

free charge carriers (electrons)

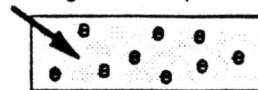

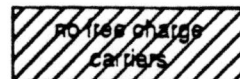

Conductor

INSULATOR: a material which has high resistivity (low conductivity).

no free charge carriers

Insulator

INTEGRATED CIRCUIT: a circuit in which many elements are fabricated and interconnected on a single chip of semiconductor material, as opposed to a "nonintegrated" circuit in which the transistors, diodes, resistors, etc. are fabricated separately and then assembled.

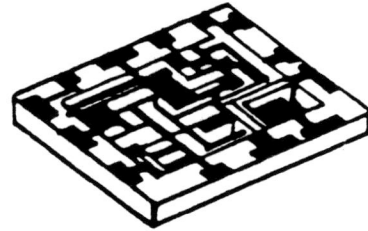

Integrated Circuit

INTEGRATOR: a sub-system of an exposure tool which detects the total energy hitting the wafer and automatically adjusts the time of exposure.

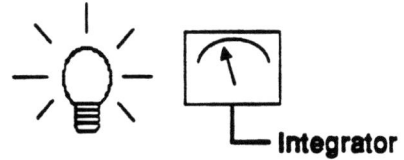

Integrator

INTERCONNECT: a thin-film conductive wiring which provides an electrical path between devices on a chip.

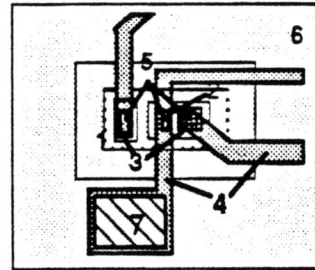

1. Buried Layer
2. Base
3. Emitter
4. **Metal to be used as Interconnects**
5. Contact
6. Passivation
7. Bonding Pad

Composite of a
Bipolar Transistor

INTRINSIC SEMICONDUC-TOR: an element or compound that has four electrons in its outer ring (i.e. elements from group IV of the periodic table or compounds of Group III and V).

Intrinsic Semiconductor
- some electrons available for conduction
- examples: Ge, Si, III-V compounds
- 10^{-9}-10^{3}ohm-cm conductivity

INTRUSION: see "Pattern Defects"

ION: an atom that has either gained or lost electrons, making it a charged particle (either negative or positive).

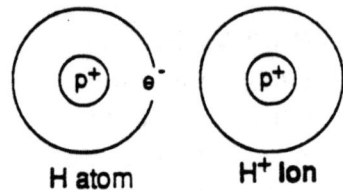

H atom   H⁺ Ion

ION BEAM MILLING: a dry etching method which uses an ion beam. Argon atoms are ionized and accelerated towards a wafer. The exposed areas are removed.

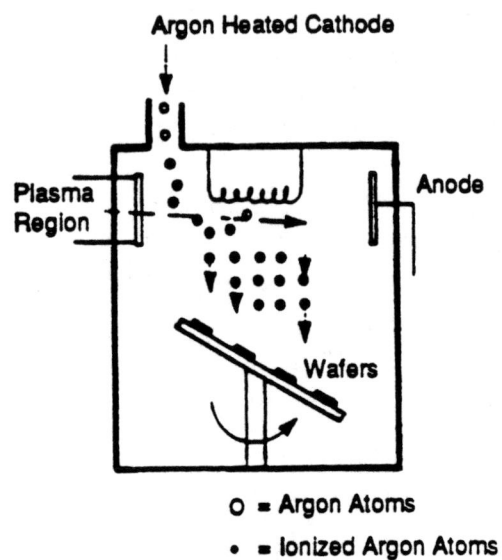

Argon Heated Cathode

Plasma Region

Anode

Wafers

o = Argon Atoms

• = Ionized Argon Atoms

Ion Milling

ION IMPLANTATION: introduction of selected impurities (dopants) by means of ion bombardment to achieve desired conductive properties in defined areas.

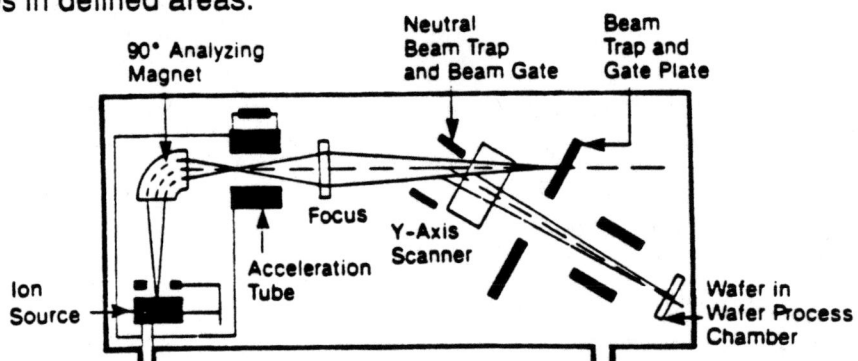

90° Analyzing Magnet

Neutral Beam Trap and Beam Gate

Beam Trap and Gate Plate

Focus

Y-Axis Scanner

Ion Source

Acceleration Tube

Wafer in Wafer Process Chamber

Ion Implanter

56

IONIZATION CHAMBER: a sub-assembly of an ion implanter in which dopant-bearing molecules are ionized by bombardment with electrons. These ions are then accelerated towards the wafer and mass analyzed. Only the desired ions are directed into the wafer.

ISLAND: patterned portion of surface layer remaining on a semiconductor wafer after a masking sequence of expose/develop/etch is completed. The island is visible in the resist after develop and is then cut into the underlying layer at the etch step.

Island

ISOLATION DIFFUSION: diffusion step resulting in P-N junctions surrounding the areas to be separated.

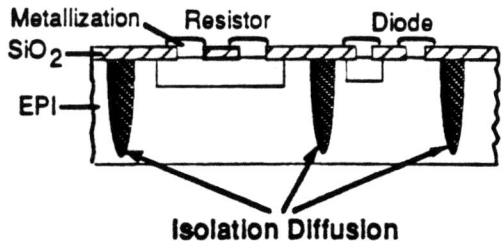

Isolation Diffusion

ISOLATION MASK: the second mask used in standard bipolar integrated circuit fabrication. Boron is diffused into the substrate in regions etched during the isolation photoresist process and serves to electrically separate (isolate) regions of the semiconducting substrate.

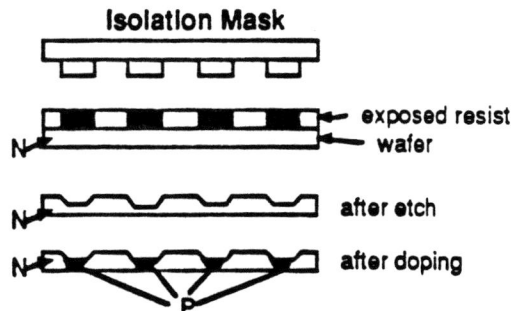

Isolation Mask

ISOPROPYL ALCOHOL (IPA): a
solvent often used in semiconductor processing for final
rinsing and drying.

IPA

ISOTROPHIC ETCHING: refers to
the etching of photoresist both
downward and to the side
resulting in pattern features
which are angled with respect
to the underlying layer.

INTERCONNECT: see "Lead".

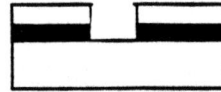

Anisotropic Etch

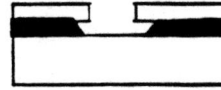

Isotropic Etch

J

JFET: see "Junction Field Effect Transistor".

JUNCTION: the interface at which the conductivity type of a material changes from P-type to N-type or vice versa.

free charge

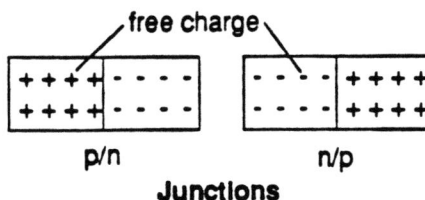

p/n   n/p

Junctions

JUNCTION DEPTH: the depth of a junction downward into the wafer.

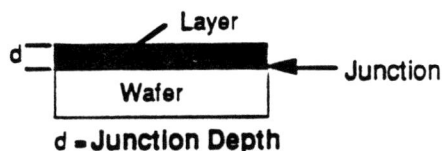

d = **Junction Depth**

JUNCTION FIELD EFFECT TRANSISTOR (JFET): a device in which voltage is applied to a terminal to control current between the source and drain regions.

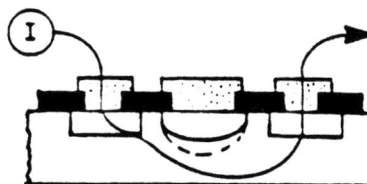

Junction Field Effect Transistor

JUNCTION TRANSISTOR: a bipolar transistor constructed from interacting P/N junctions. The term is used to distinguish junction transistors from other types, such as field-effect and point-contact transistors.

K

KV: Kilovolts (1000 volts)

KILO: prefix representing multiples of 1000, symbolized by the prefix "k".

$$1000 \ (10^3)$$
Kilo

KINEMATIC VISCOSITY: absolute viscosity divided by resist density. Measured in units of centistokes.

Kinematic Viscosity =
$$\frac{\text{Absolute Viscosity (Centipoise)}}{\text{Density (Grams/Liter)}} =$$
centistokes

L

LED: see "Light-Emitting Diode"

LSI: see "Large Scale Integration"

LASER (Light Amplification by Stimulated Emission of Radiation): a device which simplifies or generates light. Often resulting in narrow beams of high intensity pure color.

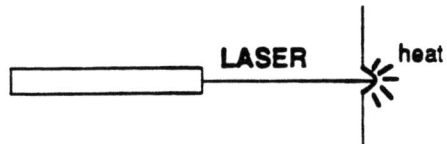

LARGE SCALE INTEGRATION (LSI): Refers to chips with between 5,000 and 100,000 components each.

Integration Levels Chart

Level	Abreviation	# Components per Chip
Small Scale Integration	SSI	2 - 50
Medium Scale Integration	MSI	50-5000
Large Scale Integration	**LSI**	**5000-100,000**
Very Large Scale Integration	VLSI	100,000-1,000,000
Ultra Large Scale Integration	ULSI	over 1,000,000

LAYERING: a process by which thin layers of different materials are grown on, or added to, the wafer surface.

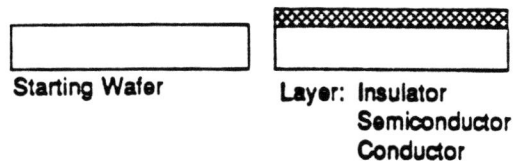

Starting Wafer   Layer: Insulator
                               Semiconductor
                               Conductor

Layering

LEAD: (1) metal component of chip package that provides an electrical path to the printed wiring board.

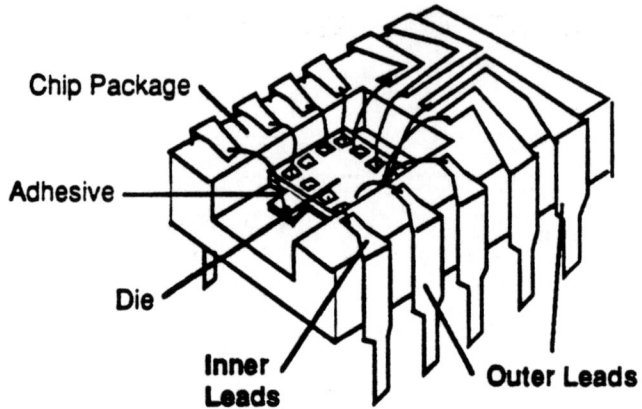

Chip Package

Adhesive

Die

Inner Leads

Outer Leads

LEAKY: a frequently used term implying the presence of an unwanted current when a voltage is applied between two points.

Leaky: undesirable current flow

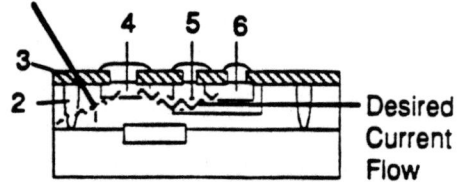

4 5 6

3

2

Desired Current Flow

1. EPI and Collector
2. Isolation
3. Surface Oxide
4. Collector Contact
5. Base
6. Emitter

Bipolar Transistor

LID SEALING: an assembly process step in which a lid is sealed over the package to protect the die.

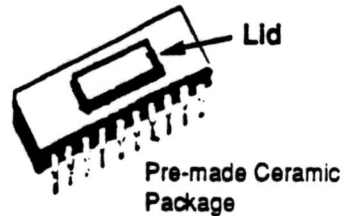

Lid

Pre-made Ceramic Package

LIGHT-EMITTING DIODE (LED): a semiconductor device in which the energy of minority carriers in combining with holes is converted to light, usually constructed as a P/N junction device.

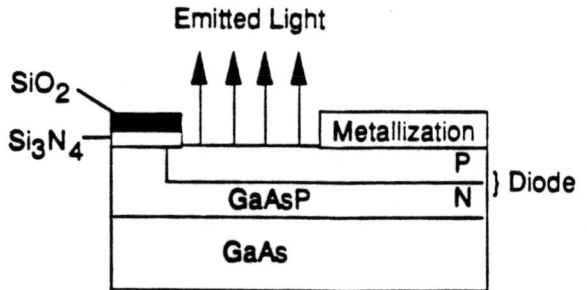

Emitted Light

SiO_2

Si_3N_4

Metallization

P

N

} Diode

GaAsP

GaAs

L.E.D. Structure

LIGHT FIELD MASK: see "Clear Field Mask".

LINE WIDTH: measurement of the shorter dimension of lines or spaces comprising the patterns formed when manufacturing devices, also referred to as critical dimension or "CD". Measurement used to maintain process control

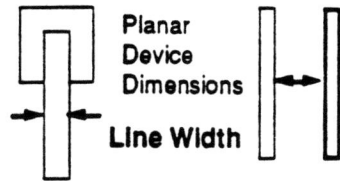

Planar Device Dimensions

Line Width

LINEAR DEVICE: see "Analog Device"

LITHOGRAPHY: process of pattern transfer. When light is utilized, it is termed photolithography. When patterns are measured in micrometers, it is referred to as microlithography.

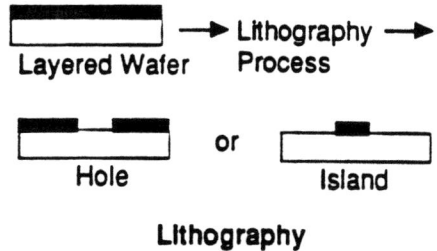

Layered Wafer → Lithography → Process

Hole   or   Island

Lithography

M

MMOS: see "Memory MOS".

MOSFET: see "Metal Oxide Semiconductor Field-Effect Transistor".

MSI: see "Medium Scale Integration".

MAJORITY CARRIER: the mobile charge carrier (hole or electron) that predominates in a semiconductor material. For example: electrons in an N-type region.

Free Charges in N-type Silicon

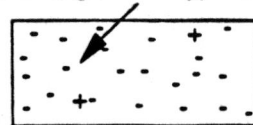

Majority Carrier: Electrons

MASK: a flat glass plate covered with an array of patterns used in the photomasking process. Each pattern consists of opaque and clear areas that respectively prevent or allow light through. Masks are aligned with existing patterns on silicon wafers and used to expose photoresist. Mask patterns may be formed in emulsion, chrome, iron oxide, silicon, or a number of other opaque materials.

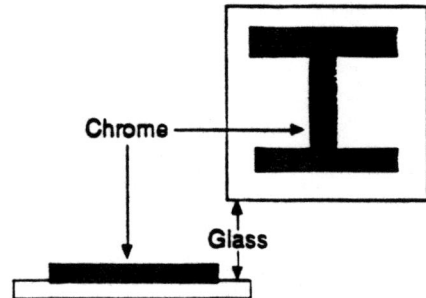

Chrome

Glass

Photo Mask

64

MASK POLARITY: a mask can be classified as either a "clear field mask" or a "dark field mask" and refers to patterns that are dominantly transparent or dominantly opaque.

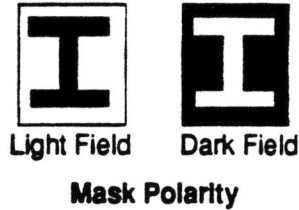

Light Field   Dark Field

Mask Polarity

MASKING: see "patterning"

MEDIUM SCALE INTEGRATION (MSI): . Refers to chips with between 50 and 5,000 components each.

Integration Levels Chart

Level	Abreviation	# Components per Chip
Small Scale Integration	SSI	2 - 50
Medium Scale Integration	**MSI**	**50-5000**
Large Scale Integration	LSI	5000-100,000
Very Large Scale Integration	VLSI	100,000-1,000,000
Ultra Large Scale Integration	ULSI	over 1,000,000

MEMORY: a circuit function which provides capacity for data or information storage.

MEMORY MOS (MMOS): A nonvolatile memory device structure that enables information to be retained during power shutdown.

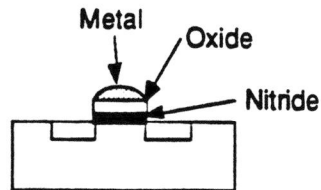

Metal   Oxide   Nitride

MMOS Structure

MESA: a device structure fabricated by selective etchings which leave flat portions of the original surface ("mesas") projecting above the neighboring regions. The mesa technique is often used to limit the extent of the electronically active material to the area of the mesa.

MESA TRANSISTOR: an early transistor based on the "Mesa" device structure.

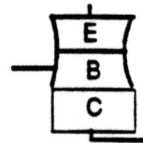

Mesa Transistor

METAL MASK: the step at which islands of conductive material are left on the wafer surface.

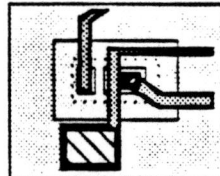

Composite of a
Bipolar Transistor

Metal Mask for
Bipolar Transitor

METAL OXIDE SEMICONDUCTOR FIELD-EFFECT TRANSISTOR (MOSFET): a device containing a metal gate over thermal oxide over silicon.

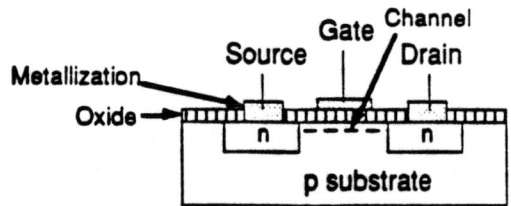

n-Channel MOSFET

METALLIZATION SEQUENCE:
a wafer with doped (electrically active) regions is patterned to create holes (contacts) down to the wafer surface, the contacts are filled with metal and a thin film of metal covers the wafer surface, this metal film is patterned to form the thin linear regions that interconnect the circuit components.

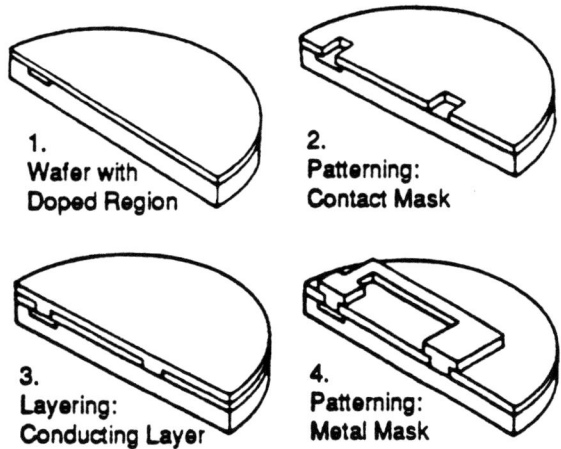

1.
Wafer with
Doped Region

2.
Patterning:
Contact Mask

3.
Layering:
Conducting Layer

4.
Patterning:
Metal Mask

Metallizaion Sequence

MICRO: refers to one-millionth. Symbol is *u(mu)*.

$$Micro = 10^{-6}$$

MICROMETER: one-millionth of a meter (10^{-6} meter), symbol is *um(mu)*.

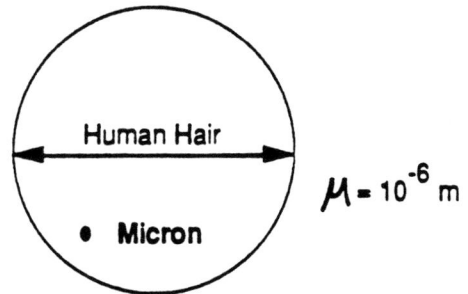

Human Hair

• **Micron**

$$\mu = 10^{-6} \text{ m}$$

MICRON: same as micrometer.

MICROCHIP: see "Chip".

MICROPROCESSOR: the computational section of a computer, referred to as MPU (micro processor unit).

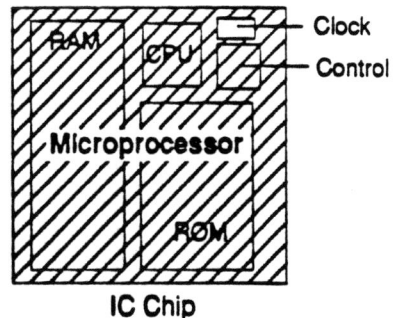

RAM

CPU

Clock

Control

Microprocessor

ROM

IC Chip

MIL: one-thousandth of an inch.
25.4 microns = 1 mil.

MILLI: one thousand. Symbolized by "m".

$$Milli = 10^{-3}$$

MILLER INDICE: a series of three numbers used to designate crystal planes on a x-y-z coordinate axis. The two most commonly used planes are <111> and <100>. The numbers indicate intersection points in the x, y, and z planes respectively.

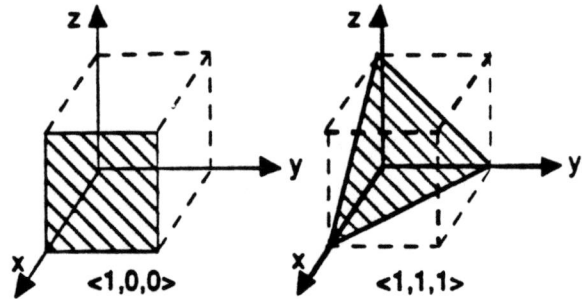

Si crystal planes
Miller Indices <1,0,0> and <1,1,1>

MINIMUM GEOMETRY: smallest line or space in a device pattern, referred to as critical dimension.

MINORITY CARRIER: the non-predominant mobile charge carrier in a semiconductor. For example: electrons in a P-type region.

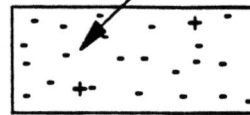

Free Charges in N-type Silicon

Minority Carrier: Holes (+)

MISALIGNMENT: occurs when a die is not correctly positioned with respect to the underlying layer

MISREGISTRATION: see "Misalignment"

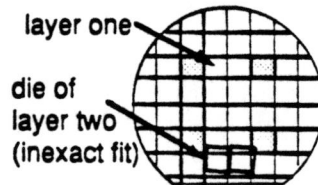

Misalignment

mm of MERCURY: a measure of pressure. 760 mm of Mercury are equivalent to atmospheric pressure (1 atm).

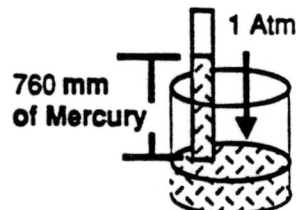

68

MOLECULE: smallest quantity of a substance that retains the properties of that substance.

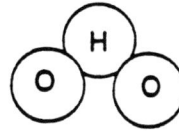

e.g. H_2O: water **Molecule**

MONOCHROMATIC LIGHT: light of a single wavelength.

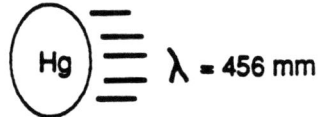

Monochromatic Light
(Example)

MONOLITHIC CHIPS: a semiconductor chip containing many devices interconnected into an integrated circuit.

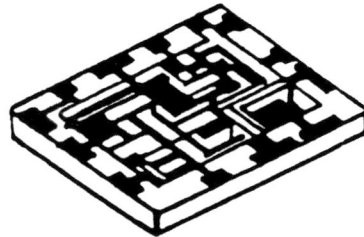

Monolithic Chip

N

NMOS: see "N-channel MOS".

NANO: one billionth, symbolized by the prefix "n".

NANOMETER: one billionth of a meter, symbolized by "nm".

N-CHANNEL MOS (NMOS): a type of MOSFET in which the channel is negative during conduction.

N-TYPE: a semiconductor material treated with a dopant having extra electrons. N-type dopants in silicon are Group V elements in which the fifth outer electron is free to conduct current in the silicon crystal.

NEGATIVE RESIST: photoresist that remains in areas that were not protected from exposure by the opaque regions of a mask while being removed by the developer in regions that were protected . A reversed image of a mask remains following the develop process. A "clear" or "light" field mask is most often used with negative resist.

$Nano = 10^{-9}$

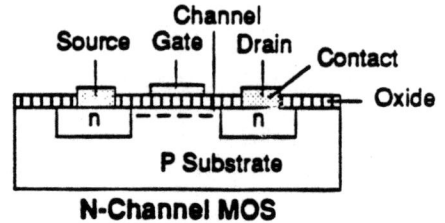

N-Channel MOS

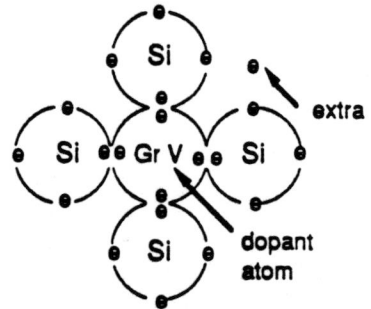

N-Type Silicon

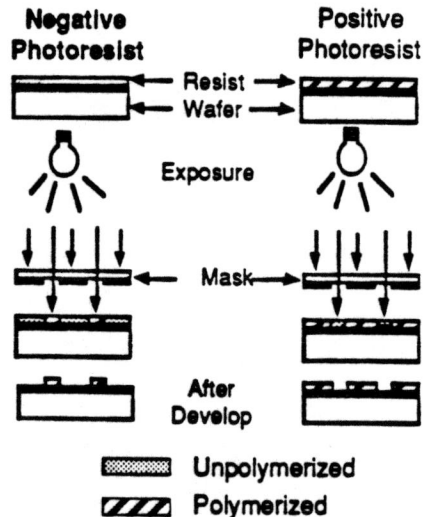

NITRIC ACID (HNO₃): a strong
acid often used to clean silicon
wafers or to etch deposited or
grown layers.

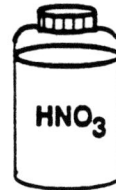

Nitric Acid

NITROGEN (N₂): an inert gas
which is often used as a carrier
gas for chemicals in semiconductor processing.

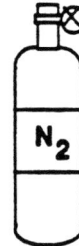

Nitrogen

NPN TRANSISTOR: a transistor
which has a base of P-type
silicon sandwiched between an
emitter and a collector of N-type
silicon.

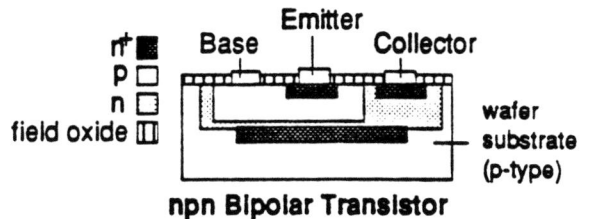

npn Bipolar Transistor

O

OHM'S LAW: a relationship between resistance, voltage and current, $V=IR$

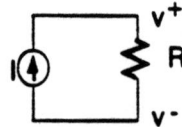

Ohm's Law: $V = I * R$
V = Voltage
I = Current
R = Resistance

OVERALL YIELD: the percentage of functioning packaged chips from a wafer compared to the number of die mapped onto the wafer at the start of processing. Overall Yield is the product of Fab Yield, Sort Yield, and Assembly Yield.

(Fab Yield) x (Sort Yield) x

(Assembly & Test Yeild) = **Overall Yield**

OXIDATION: the growth of oxide on silicon when exposed to oxygen. This process is highly temperature dependent.

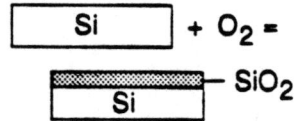

$$Si + O_2 =$$

$$Si \quad - SiO_2$$

Oxidation

OXIDATION REACTION CHAMBER: an enclosure in which oxidation takes place. It is made up of sections which service four or more vertically stacked tube furnaces.

Oxidation Reaction Chamber

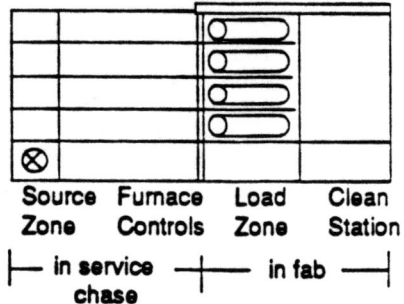

Source Zone	Furnace Controls	Load Zone	Clean Station

— in service chase — + — in fab —

OXIDE: see "Silicon Dioxide"

OXIDE ETCHING: an etching process which uses acid (usually hydrofluoric acid HF). The acid is buffered in order for the reaction to proceed at a rate slow enough to be controlled. Buffered oxide etch (BOE) is often used.

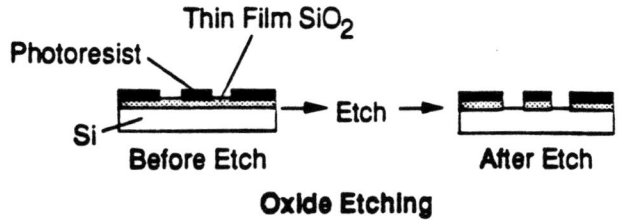

Oxide Etching

OXIDE EVALUATION: the determination of oxide thickness, dielectric strength, and index of refraction for oxide layer steps. Unpatterned test wafers are added to each boat for measuring these properties. Oxide thickness is measured after every run. Dielectric strength is determined by measuring the rupture voltage of the oxide which is a destructive test performed on the test wafer. The index of refraction is measured on a sample basis on the test wafers.

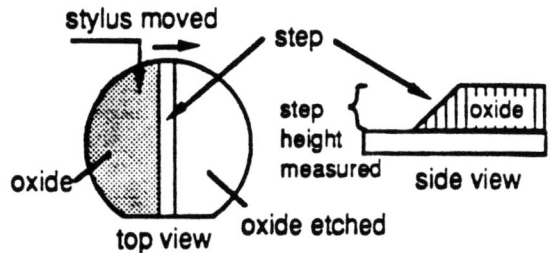

Oxide Thickness Determined

Oxide Evaluation

OXIDE MASKING: the use of an oxide layer (on a semi-conducting substrate) in which a pattern is formed. The pattern is used to define the location for diffusion or implantation of impurities (dopants).

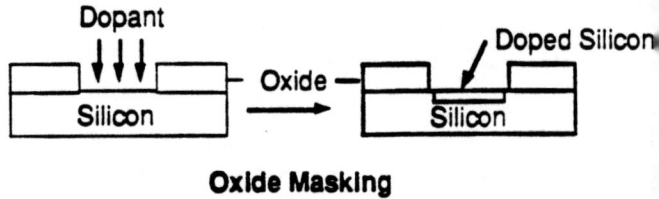

Oxide Masking

OXYGEN (O_2): a gas used to combine with silicon to form silicon dioxide. Also used to remove resist from wafers in dry removal processes.

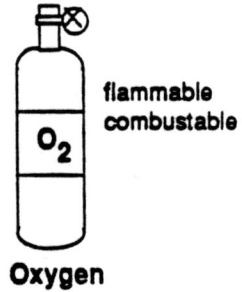

Oxygen

74

P

PAL: see "Programmable Array Logic".

PMOS: see "P-channel MOS".

PVX: see "Phosphorus-Doped Vapor-Deposited Oxide".

PROM: see "Programmable Read Only Memory".

P-CHANNEL MOS (PMOS): a type of MOSFET in which the source and drain are made of P-type materials and the channel allowing current flow is P-type. Thus, conduction is achieved by movement of holes.

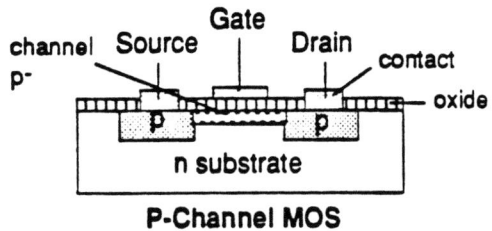

P-Channel MOS

P-TYPE: a semiconductor material treated with a dopant which has one less electron than silicon. P-type dopants in silicon are Group III elements in which the missing electron leaves a hole which is free to conduct charge.

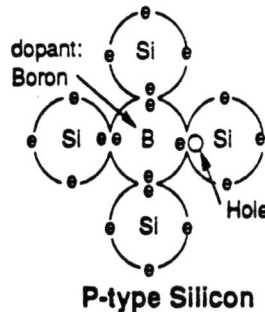

P-type Silicon

PACKAGE: a protective container for a semiconductor chip which

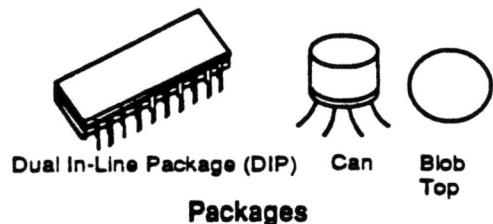

Dual In-Line Package (DIP)   Can   Blob Top

Packages

PASSIVATION: sealing layer, usually silicon dioxide or silicon nitride, added at the end of the fabrication process to prevent deterioration of electronic properties through chemical action, corrosion, or handling during the packaging processes. Continues to protect packaged chips against moisture and contamination.

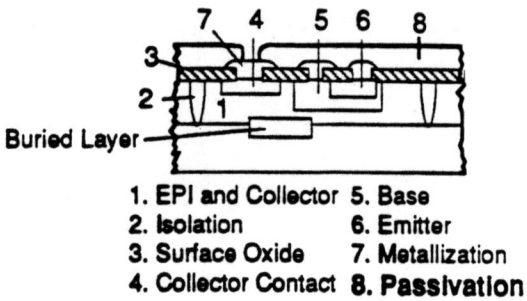

1. EPI and Collector 5. Base
2. Isolation   6. Emitter
3. Surface Oxide   7. Metallization
4. Collector Contact 8. **Passivation**

PASSIVATION SEQUENCE: a sequence in which a doped and metallized chip is patterned in order to cover all but the bonding pads. This layer acts as a contamination barrier and protects the thin metal layer against scratching.

PATTERN DEFECTS: process induced feature anomalies on masks and wafers.

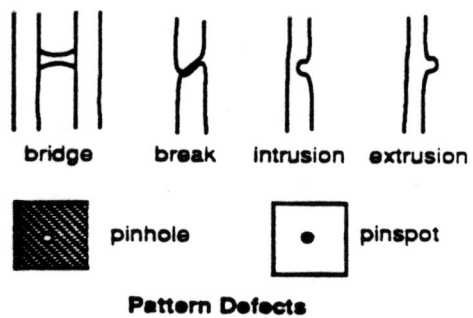

bridge   break   intrusion   extrusion

pinhole   pinspot

Pattern Defects

PATTERN GENERATOR: a
machine comprised of a light
source and computer controlled
high speed shutter. A reticle
blank coated with photoresist is
placed on a stage and is moved
under the shutter and light
source systems to expose the
reticle in the same pattern as
the original digitized drawing.

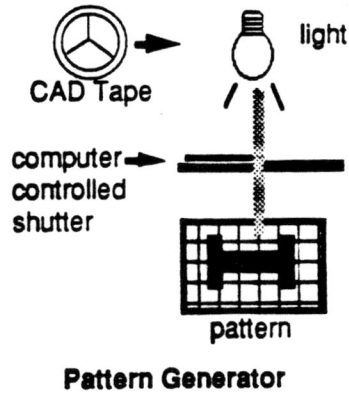

Pattern Generator

PATTERNING: method used to
create features consisting of
lines and spaces.

PELLICLE: a thin film of an optical
grade polymer that is stretched
on a frame and secured to a
mask or reticle. This solves the
problem of airborne dirt collect-
ing on the mask and acting as
an opaque spot. The image of
the contamination is out of the
focal plane during exposure and
does not print on the wafer.

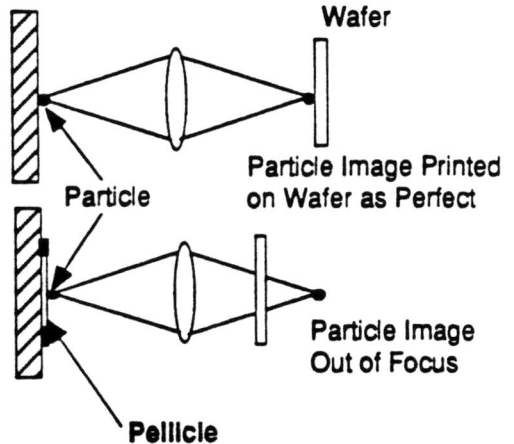

PERIODIC TABLE: the table of
chemical elements arranged in
rows and columns. Elements
with similar properties are in the
same column.

Periodic Table

PHOSPHINE (PH$_3$): a gas that is often used as a source of phosphorus for doping silicon.

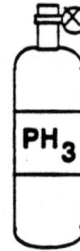

Phosphine

PHOSPHORUS (P): the N-type dopant commonly used for the sink and emitter diffusions in standard bipolar integrated circuit technology.

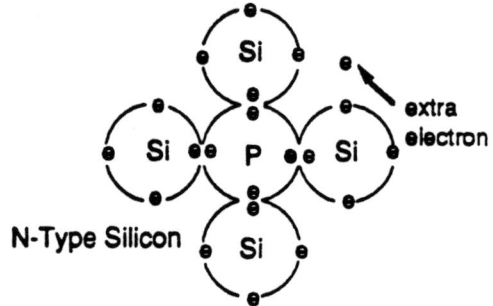

N-Type Silicon

Phosphorus

PHOSPHORUS-DOPED VAPOR-DEPOSITED OXIDE (PVX): (also known as Doped Silox) a chemically deposited layer of phosphorus-rich silicon dioxide. PVX can be used for scratch protection. PVX is often used with a layer of vapox.

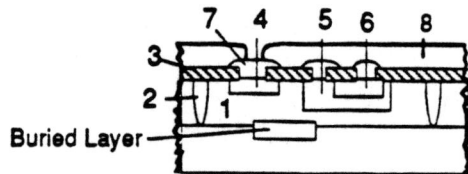

Buried Layer

1. EPI and Collector   5. Base
2. Isolation   6. Emitter
3. Surface Oxide   7. Metalization
4. Collector Contact   8. Glassivation:
                           e.g. PVX

PHOSPHORUS OXYCHLORIDE (POCl$_3$): a liquid that is often used as a source of phosphorus for doping silicon.

78

PHOTODIODE: a diode sensitive to light in which the current is proportional to the intensity of the incoming light.

PHOTOLITHOGRAPHY: a process in which the pattern in a reticle or photomask is transferred to a wafer resulting in formation of a pattern that identifies areas to be doped or selectively removed.

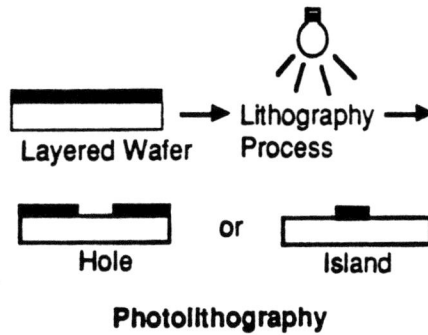

Photolithography

PHOTOMASKING: see "Photolithography".

PHOTOPLATE: term used to describe a coated mask blank.

Photoplate

PHOTORESIST: the light-sensitive material spun onto wafers in a thin film and "exposed" using high intensity light projected through a mask. The exposed (or unexposed photoresist depending on the polarity of the photoresist) photoresist is dissolved with developers, leaving a pattern which allows etching to take place in some areas while preventing it in others.

PHOTORESIST POLARITY: photoresists can be positive or negative. see "positive resist" or "negative resist".

PHOTOSOLUBILIZATION: the reaction in which photoresist resin changes from an insoluble to a soluble state.

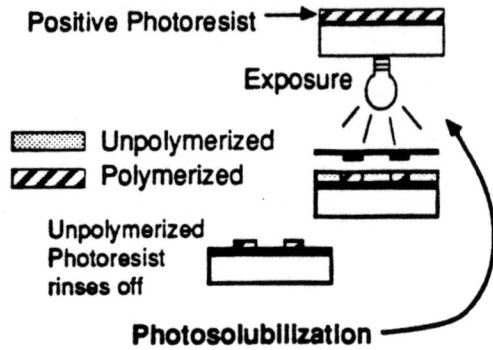

Positive Photoresist

Exposure

Unpolymerized
Polymerized

Unpolymerized Photoresist rinses off

Photosolubilization

PICK: a pre-assembly step in which the non-inked die (electrically functioning) are selected and sent into the assembly operation.

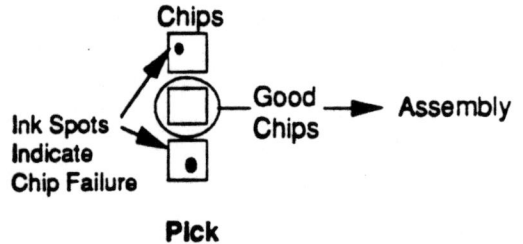

Chips

Ink Spots Indicate Chip Failure

Good Chips

Assembly

Pick

PIN GRID ARRAY: a ceramic chip package designed for larger chips. This package has more leads available than in a DIP.

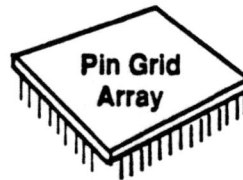

Pin Grid Array

PINHOLE: a small undesired hole in a photoresist layer or in the opaque region of a mask or reticle. A pinhole in a mask or reticle results in the formation of a hole in imaged photoresist.

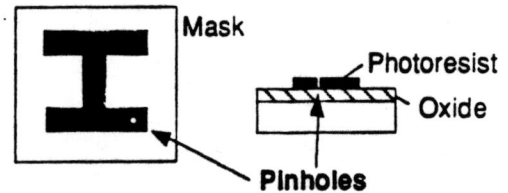

Mask

Photoresist
Oxide

Pinholes

PLANAR STRUCTURE: a flat-surfaced device structure fabricated by diffusion and oxide masking with junctions terminating in a single plane.

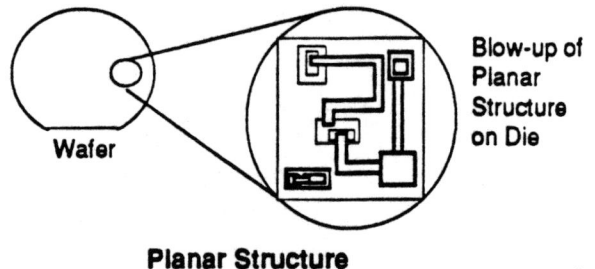

Wafer

Blow-up of Planar Structure on Die

Planar Structure

PLANETARY FIXTURE: a mechanism which rotates wafers enabling the wafer surface to be evenly coated by an evaporation source.

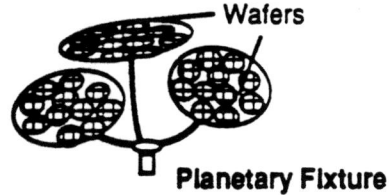

Planetary Fixture

PLASMA: high energy gas made up of ionized particles.

PLASMA ETCHER: a machine in which a high energy RF field excites the gas molecules in the chamber to a high level causing a reaction in which unprotected sections of an oxide layer are removed.

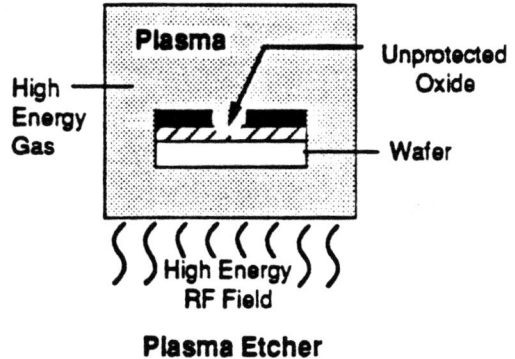

Plasma Etcher

PLASMA ETCHING: an etching process which accomplishes results similar to the chemical etch mechanism reaction using an etching gas instead of a wet chemical.

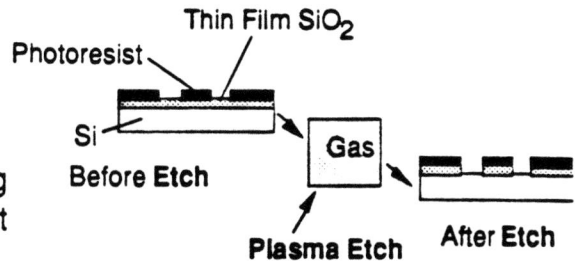

P/N JUNCTION: an interface-within a crystal between a P region which conducts primarily by holes and an N region which conducts primarily by electrons.

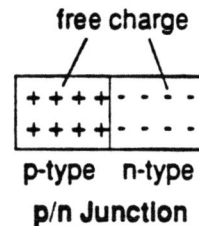

p/n Junction

PNP: semiconductor crystal structure consisting of an N-type region sandwiched between two P-type regions. Commonly used in bipolar transistors.

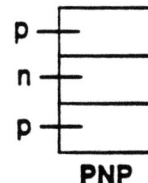

PNP

81

POLYMER: a complex chemical compound made up of repeating units.

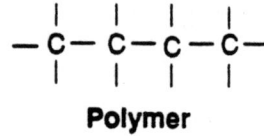
Polymer

POLYMERIZATION: the formation of crosslinked chains among molecules. This process occurs when negative photoresist responds to light. A substance which is insoluble in negative developer is formed resulting in a protective etch barrier on the wafer surface.

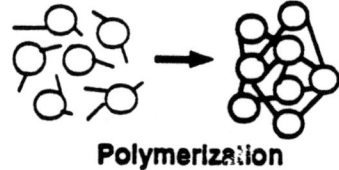
Polymerization

POLYCRYSTALLINE SILICON (Poly): silicon composed of many crystals randomly arranged.

POLYSILICON: see "Polycrystalline Silicon"

Crystalline
Polycrystalline

POSITIVE RESIST: photoresist which becomes removable when exposed to light. Photoresist areas protected from exposure by the opaque regions of a mask remain after develop. A positive (similar) image of the mask results. A "dark field" mask is used most often with positive resist. Positive resist is generally used for pattern features smaller than 3 microns.

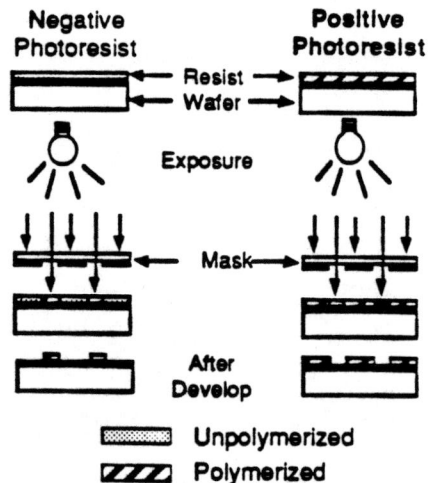

Negative Photoresist   Positive Photoresist

Resist
Wafer
Exposure
Mask
After Develop

▨ Unpolymerized
▨ Polymerized

POWER: electrical energy expressed in units of "watts".

Power = current x voltage

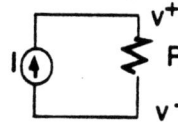

P(dissipated by resistor: R) = I x V

PREDEPOSITION (Predep): the process step during which a controlled amount of a dopant is introduced into the crystal structure of a semiconductor.

PRIMER CHEMICAL: a chemical which enhances the adhesion of a desired layer. In semiconductor technology, the desired layer is usually photoresist and a common primer chemical is HMDS.

Primer Chemical
usually **HMDS**

PRINTED CIRCUIT BOARD: a "plastic" board onto which chips are attached. Conductive metal paths on the surface of the board connect the various chips.

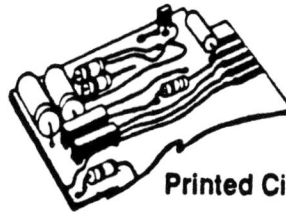

Printed Circuit Board

PROBING: a test technique used after the completion of the wafer fabrication process to evaluate each die electrically. "Probes" are positioned on the bonding pads of each die to conduct the tests.

Probing

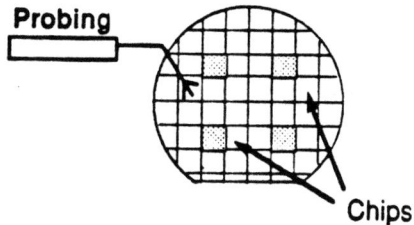

Chips

PROGRAMMABLE ARRAY LOGIC (PAL): A logic device which may be programmed by the user rather than custom-made by the chip manufacturer.

PROGRAMMABLE READ ONLY MEMORY: A technology in which fuses are used in every memory cell and selected fuses are blown in order to program the chip with user-specified information.

PROJECTION ALIGNMENT: an exposure system in which the image on the mask is projected onto the wafer. Projection alignment is commonly utilized for LSI and non-critical layers of VLSI products.

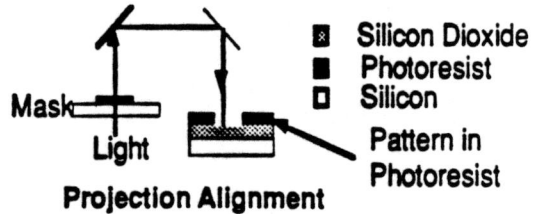

▨ Silicon Dioxide
■ Photoresist
◻ Silicon

Pattern in Photoresist

Projection Alignment

PROXIMITY EXPOSURE: this exposure method positions the wafer close to, but not touching, the mask. It is limited in image dimensional control (5 microns) due to the diffraction of light around the mask.

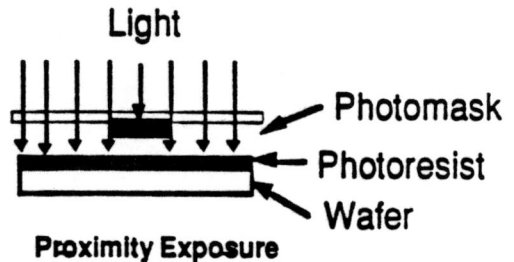

Light

Photomask
Photoresist
Wafer

Proximity Exposure

PYROLYZE: decompose.

Q

QUARTZ: commercial name for silicon dioxide formed into glass products. Because of its high temperature resistance, quartz fixtures are used in many processing steps.

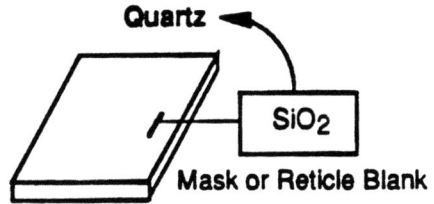

Quartz → SiO_2

Mask or Reticle Blank

QUALIFICATION: evaluation of production equipment, materials, or processes.

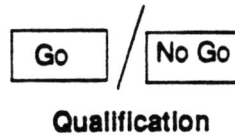

Go / No Go

Qualification

R

RAM: see "Random Access Memory".

RF: radio frequency.

RIE: see "Reactive Ion Etch"

RF HEATING: see "Radio Frequency Heating".

ROM: see "Read Only Memory".

RADIATION: electromagnetic waves that travel through the air as well as through a vacuum. Electromagnetic waves carry energy which is transferred directly to the solid objects they hit. Radiation is one of the three methods of heat transfer (conduction, convection and radiation).

Radiation ΔH

RADIO FREQUENCY (RF) HEATING: a type of heating used to achieve the 1425 C melting temperature of silicon.

RANDOM ACCESS MEMORY (RAM): a device that temporarily stores digital information. Information can be read from the device or written to it.

RASTER SCANNING: movement from left to right followed by right to left (or reversed). Used to direct an E-beam to the location where the pattern is to be written. During wafer scanning, the beam is activated only at locations requiring exposure.

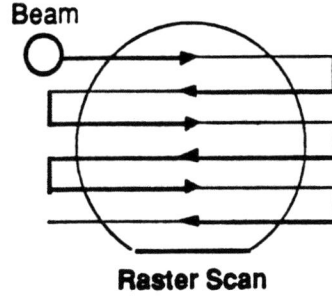

Raster Scan

RCA CLEAN: a multiple step process to clean wafers before oxidation. Named after RCA, the company that developed the procedure.

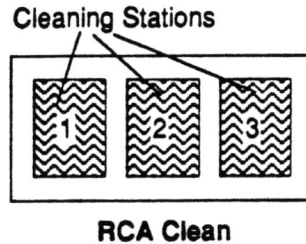

Cleaning Stations

RCA Clean

REACTANT: in a chemical reaction, a chemical which reacts with others to form products.

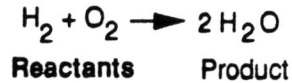

$$H_2 + O_2 \rightarrow 2H_2O$$

Reactants   Product

REACTANTS SOURCE CABINET: a cabinet at the back end of an oxidation/diffusion furnace which contains a gas flow controller and oxidant/dopant source.

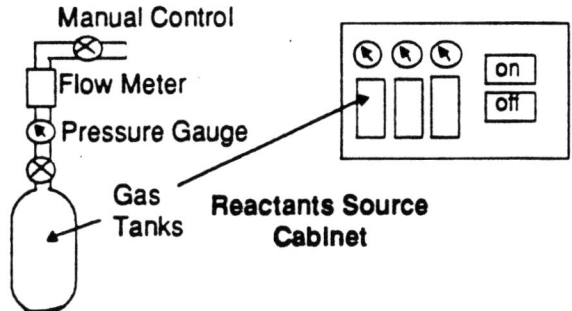

Manual Control
Flow Meter
Pressure Gauge
Gas Tanks

Reactants Source Cabinet

READ ONLY MEMORY (ROM): a device in which information is permanently stored and cannot be changed.

REATIVE ION ETCHING (RIE):
an etching process that com-
bines plasma and ion beam
removal of the surface layer.
The etchant gas enters the
reaction chamber and is ionized.
The individual molecules accel-
erate to the wafer surface. At
the surface, top layer removal is
achieved by the physical and
chemical action on the material.

REACTOR: chamber used for the
deposition or removal of layer
material used in semiconductor
processing. Common types of
reactors are epitaxial reactors,
vapox reactors, nitride reactors,
and plasma etchers.

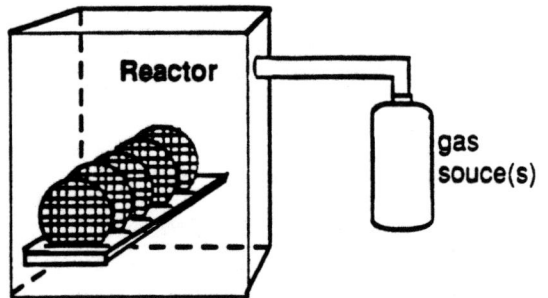

REGISTRATION: the positioning
of patterns relative to patterns
previously created in or on the
wafers' surface. Same as alignment.

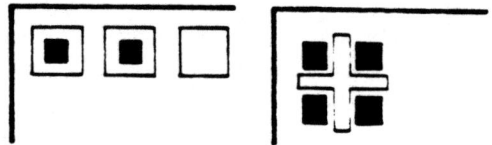

Registration Marks

RESISTANCE: opposition to
current flow.

RESISTIVITY: a measure of the resistance to current flow in a material. The more tightly bound the electrons of a material are, the greater the resistivity.

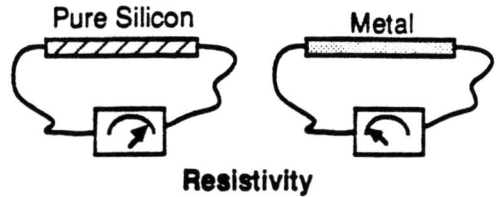

Pure Silicon   Metal

Resistivity

RESISTOR: a device or circuit element which dissipates power.

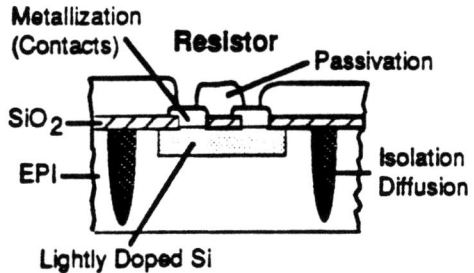

Metallization (Contacts)   **Resistor**   Passivation
SiO$_2$
EPI
Isolation Diffusion
Lightly Doped Si

RESOLUTION: the smallest opening or space that can be clearly imaged or viewed.

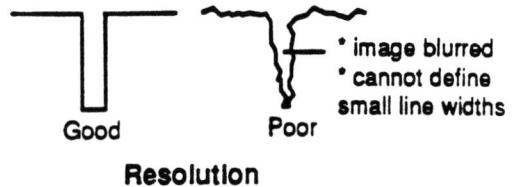

* image blurred
* cannot define small line widths

Good   Poor

Resolution

RETICLE: a reproduction of the pattern to be imaged on the wafer (or mask) by a step and repeat process. The actual size of the pattern on the reticle is usually several times the final size of the pattern on the wafer. The reticle consists of an emulsion or chrome photo plate that is selectively exposed to light by a pattern generator or has its pattern created by an E-Beam machine.

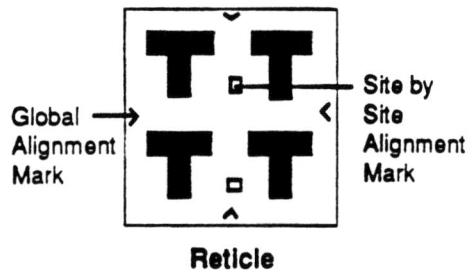

Global Alignment Mark   Site by Site Alignment Mark

Reticle

REWORK: rejected wafers from develop inspection that are sent through the photomasking process again.

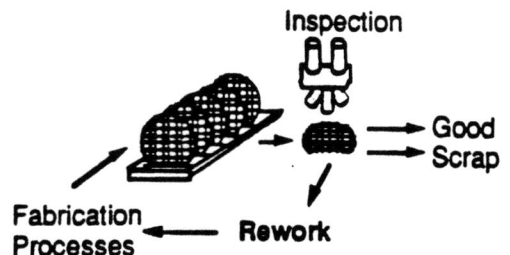

Inspection

Good
Scrap

Fabrication Processes   **Rework**

89

REWORK RATE: the percentage of wafers requiring rework.

Rework Rate =

$$\frac{\text{\# Reworked Wafers}}{\text{\# Wafers Started into Fab}} \times 100\%$$

RINSE: the removal of wet etchants or developers from the wafer. This process results in stopping the etching or developing processes by removing the active chemical from the surface. There are several different methods of rinsing: overflow rinsing, spray rinsing, dump rinsing, and spin/rinse drying.

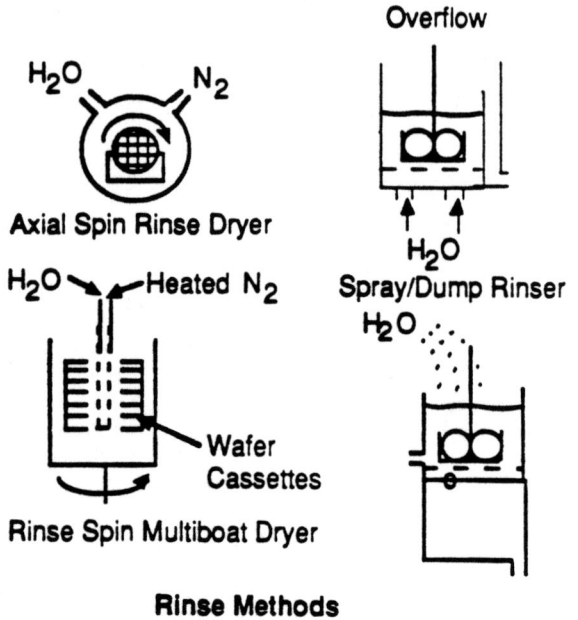

H_2O N_2

Axial Spin Rinse Dryer

H_2O Heated N_2

Wafer Cassettes

Rinse Spin Multiboat Dryer

Overflow

H_2O

Spray/Dump Rinser

H_2O

Rinse Methods

RUPTURE VOLTAGE: the voltage at which a dielectric (an insulator) breaks down and conducts charge.

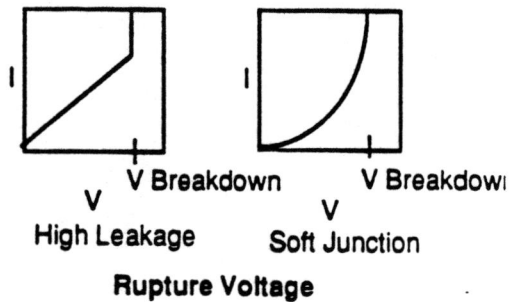

I

V Breakdown

V

High Leakage

I

V Breakdown

V

Soft Junction

Rupture Voltage

S

SEM: see "Scanning Electron Microscope"

SIC: see "Silicon Integrated Circuit".

S.O.T.A.: State of the Art., used to describe leading edge technology.

SSI: see "Small Scale Integration".

SCALING: a method of decreasing line and space pattern sizes during the imaging process.

SCANNING ELECTRON MICROSCOPE (SEM): a microscope used to magnify images as much as 50,000 times by means of scanning with an electron beam. The scanning electrons cause electrons on the surface to be ejected. The ejected electrons are collected and translated into a picture of the surface.

SCHEMATIC DIAGRAM: a circuit diagram using symbols and line connections to represent actual physical components and interconnects in the circuit.

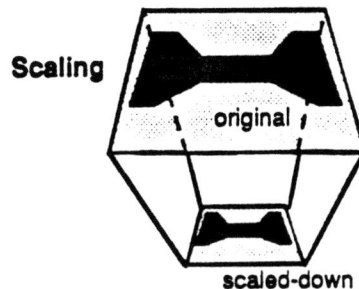

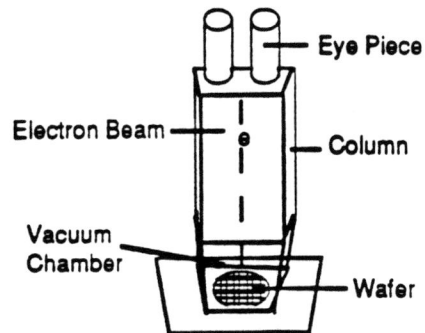

Scaling

original

scaled-down

Eye Piece

Electron Beam

Column

Vacuum Chamber

Wafer

Scanning Electron Microscope (SEM)

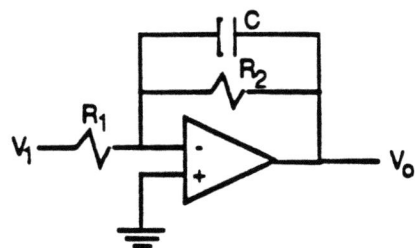

C

R_2

V_1 — R_1

V_o

Schematic Diagram

SCHOTTKY DIODE: a diode
which allows a lower voltage
drop than an ordinary diode and
has a faster on/off switching
speed.

SCRIBE LINES: lines separating
die on a wafer. The wafer will
be sawed along the scribe lines
resulting in individual chips.

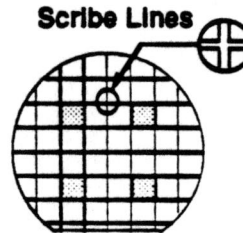

Scribe Lines

SEMI: Semiconductor Equipment
and Materials International. A
trade organization responsible
for formulation of many industry
guidelines as well as sponsor-
ship of the main industry trade
shows (Semicon).

S E M I

SEMICONDUCTOR: an element such as silicon or germanium, intermediate in electrical conductivity between the conductors and the insulators. Conduction takes place by means of holes and electrons. In the vernacular, used to refer to integrated circuits.

Electrical Classification

Classification	Examples	Conductivity	Electrons
1.Semiconductor			
a. Intrinsic	Germanium Silicon III-V	10^{-9} -10^3 /ohm-cm	Some available
b. Doped	N-type P-type		Controlled amount available
2. Conductor	Gold Silver	10^4-10^6 /ohm-cm	Free to move
3 Insulator (Dielectric)	Glass Plastic	10^{-22} -10^{-10} /ohm-cm	Bound

SEMICONDUCTOR GRADE SILICON: see "Electrical Grade Silicon".

SENSITIZER: 1) in negative photoresist: a chemical added to a photoresist to either increase or decrease the range of wave lengths to which the photoresist responds. 2) in positive photoresist: a chemical added to change the film from solvent soluble to water soluble.

SHEET RESISTANCE: a measurement of resistance with dimensions of ohms/ cm^2 that shows the number of N-type or P-type donor atoms in a semiconductor.

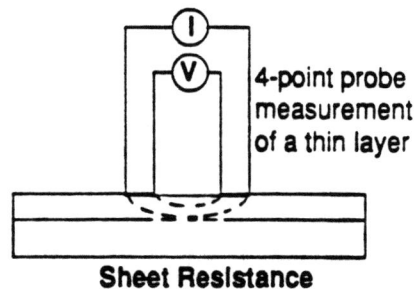

4-point probe measurement of a thin layer

Sheet Resistance

SILICON (SI): common semi-conducting material used for fabricating diodes, transistors and integrated circuits.

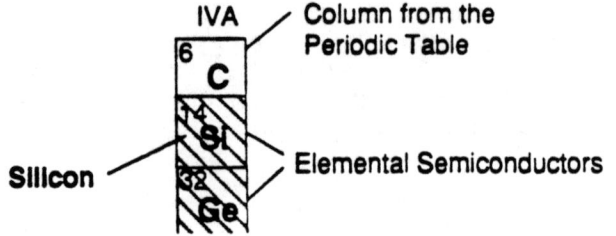

SILICON DIOXIDE (SIO$_2$): a non-conducting layer that can be thermally grown or deposited on silicon wafers. Thermal silicon dioxide is commonly grown using either oxygen or water vapor at temperatures above 900 C.

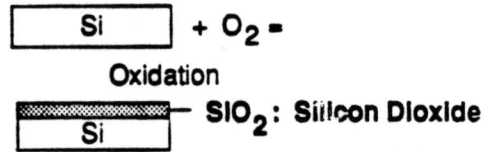

SILICON FOUNDRY: a custom circuit supplier. A manufacturer which produces chips from outside supplied designs and masks and/or produces master chips that contain standard device functions. The customer can specify the desired function and the foundry will "wire in" the function at the metallization step.

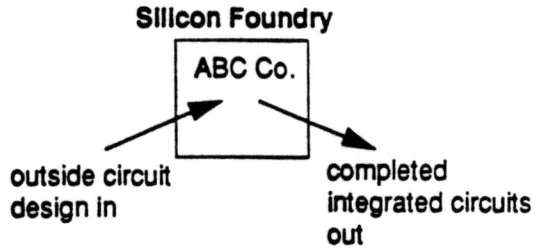

Silicon Foundry

outside circuit design in

completed integrated circuits out

SILICON INTEGRATED CIRCUIT (SIC): an integrated circuit where all of the components such as transistors, diodes, resistors and capacitors are successively fabricated in or on the silicon and interconnected.

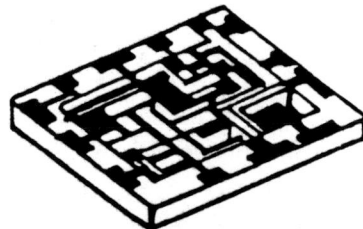

Silicon Integrated Circuit

94

SILICON NITRIDE (Si$_3$N$_4$): a non-conductive layer chemically deposited on wafers at temperatures between 600 and 900 C. When it is the final layer in the fabrication process, it protects devices against contamination.

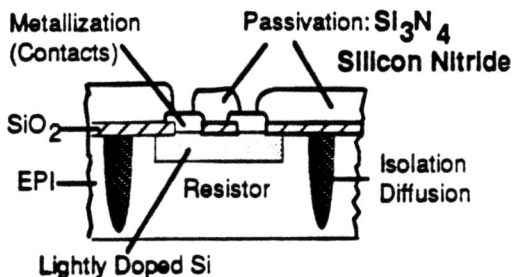

Metallization (Contacts)

Passivation: Si$_3$N$_4$ Silicon Nitride

SiO$_2$

EPI

Resistor

Isolation Diffusion

Lightly Doped Si

SINGLE CRYSTAL: refers to substances which have all unit cells arranged in a definite and repeated fashion as opposed to polycrystalline materials in which the unit cells are randomly arranged.

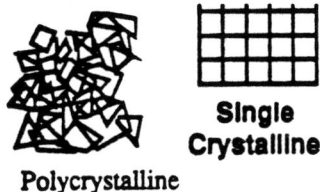

Single Crystalline

Polycrystalline

SLIP: a crystal growth defect that occurs when the crystal groups of atoms slip or shift along crystal planes.

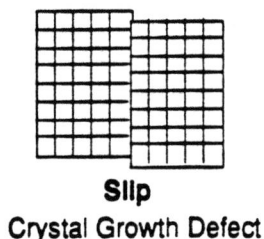

Slip
Crystal Growth Defect

SLOPE ETCHING: controlled undercutting. An etch strategy in which the sides of the contact holes are purposely overetched so as to reduce the shadow effect of the side wall and the resultant thinning of the film applied in a subsequent step.

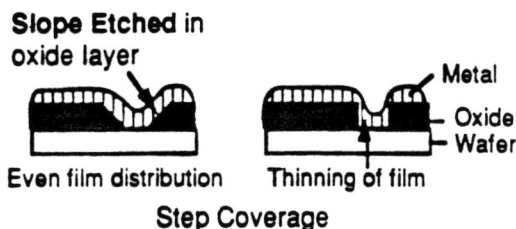

Slope Etched in oxide layer

Metal
Oxide
Wafer

Even film distribution

Thinning of film

Step Coverage

SLUG: see "Buried Layer".

**SMALL SCALE INTEGRA-
TION (SSI):** refers to
chips with between 2
and 50 components
each.

Level	Abreviation	# Components per Chip
Small Scale Integration	SSI	2 - 50
Medium Scale Integration	MSI	50-5000
Large Scale Integration	LSI	5000-100,000
Very Large Scale Integration	VLSI	100,000-1,000,000
Ultra Large Scale Integration	ULSI	over 1,000,000

SOFT BAKING: a heating
process used to evapo-
rate a portion of the
solvents in resist. The
term "soft" describes the
still soft resist after baking. The solvents are
evaporated to achieve
two results: to avoid re-
tention of the solvent in
the resist film and to
increase the surface ad-
hesion of the resist to
the wafer.

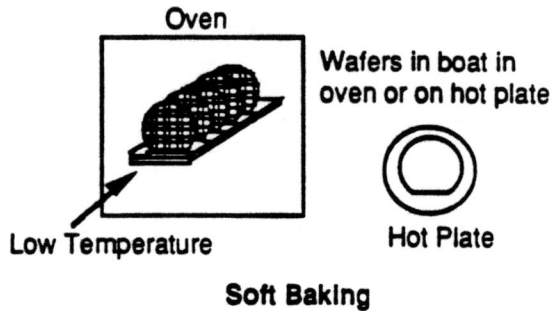

Oven

Wafers in boat in
oven or on hot plate

Low Temperature

Hot Plate

Soft Baking

SOFTWARE: computer
programs and codes
that provide instructions
to computers.

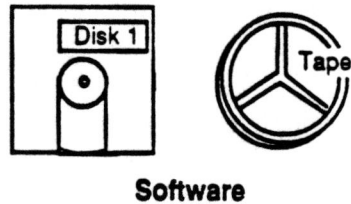

Disk 1

Tape

Software

SOLAR CELL: a large-
area diode in which a P/
N junction close to the
surface of a semicon-
ductor generates electri-
cal energy from light
falling on the surface.

SOLID STATE: designation used to describe devices and circuits fabricated from solid materials such as semiconductors, ferrites, or thin films, as distinct from devices and circuits making use of electron tube technology.

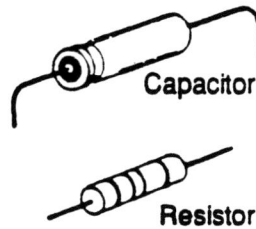

Solid State Electronics

SOLUBLE: able to be dissolved.

SOLVENT: a liquid that serves as an application vehicle for the photosensitive polymers used in photoresist. The solvent has the same function as the liquid base of a paint.

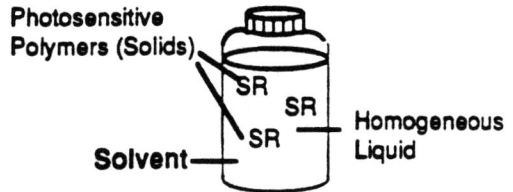

SOURCE: one of the three regions of a unipolar or field-effect transistor (FET), along with the gate and drain. Current flows between the source and drain as controlled by the gate.

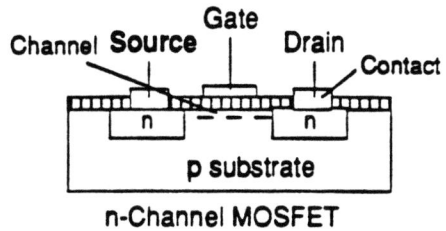

n-Channel MOSFET

SPECTRAL RESPONSE: the range of wavelengths to which a photoresist responds as determined by the sensitizers which it contains.

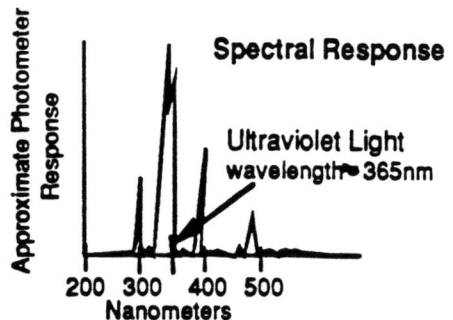

SPECTROPHOTOMETER:
an analytical instrument
used to collect interfer-
ence measurements
which are used to calcu-
lates film thickness

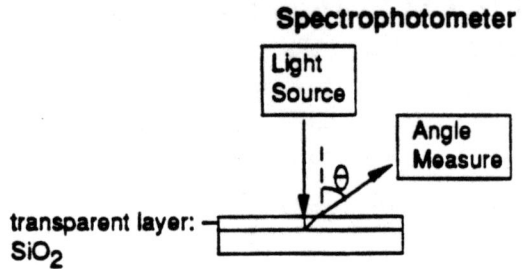
Spectrophotometer

SPIN COATING: see "Spinning"

SPINNING: a technique in
which the photoresist is
coated on the wafer
surface resulting in a
typical photoresist layer
0.5 microns thick with an
allowable thickness vari-
ation of 10%.

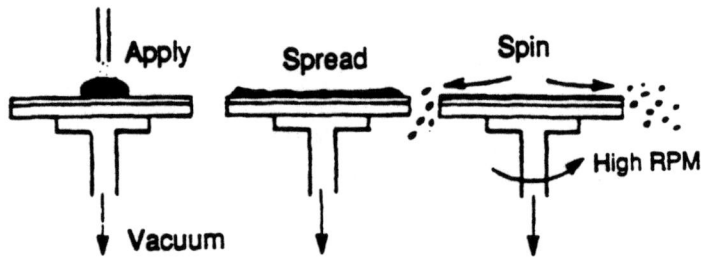
Photoresist Application by Spinning

SPREADING RESISTANCE:
a technique used for
measuring the dopant
concentration profile in a
wafer.

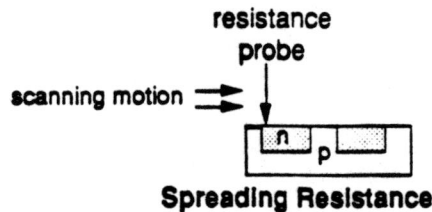
Spreading Resistance

SPUTTERING: a method of
depositing a thin film of
material on wafer sur-
faces. A target of the
desired material is bom-
barded with RF-excited
ions which knock atoms
from the target. The
dislodged target material
deposits on the wafer surface.

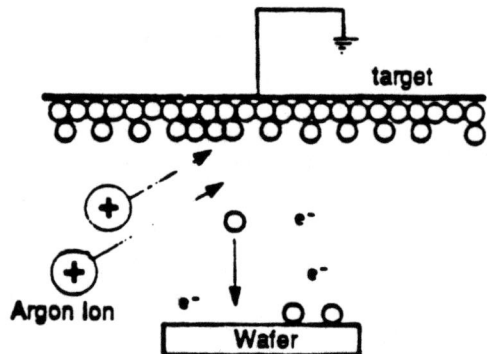
Sputtering

STAINED PADS: see "Brown Pads"

STANDARD CELLS: redundant logic elements that may be selected and grouped to create a custom IC

STATIC CHARGE: electric charge which is a source of contamination in the semiconductor manufacturing process causing wafers to pick up unacceptable levels of particulates from the air and work surfaces.

STATIC RAM: Static Random Access Memory device, retains information permanently. A Static RAM does not need a refresh function to maintain data as does a Dynamic RAM.

STEAM OXIDE: thermal silicon dioxide grown by bubbling a gas (usually oxygen or nitrogen) through water at 98 - 100 C.

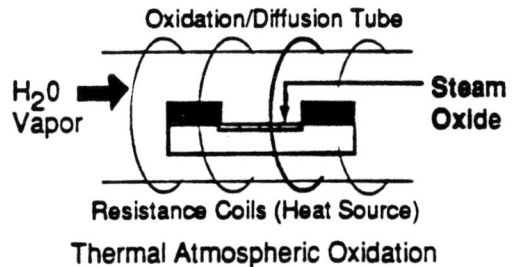

Thermal Atmospheric Oxidation

STEP AND REPEAT: an operation in which the pattern on a reticle is transferred to the mask or wafer. The photoresist coated mask blank (chrome, emulsion, or iron oxide) or wafer is placed on an X-Y stage and the reticle pattern is repeatedly imaged until the entire surface is filled with the reticle pattern.

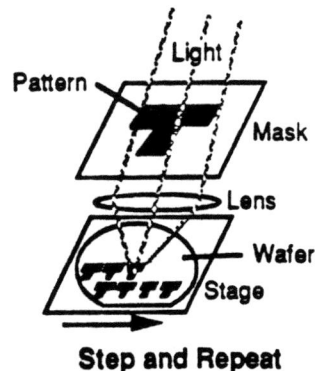

Step and Repeat

99

STEP COVERAGE: the ability of new layers to evenly cover steps formed in the existing wafer layers.

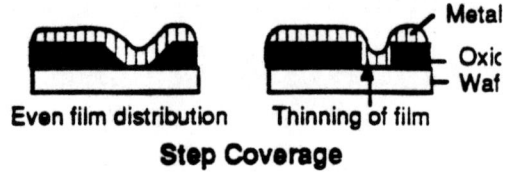

Even film distribution   Thinning of film

Step Coverage

STEPPER: a machine which steps a reticle directly onto the wafer. Alignment of reticle to wafer is accomplished by reflecting a laser beam through a special reticle pattern (alignment target) and off a corresponding pattern on the wafer assuring registration at each die location.

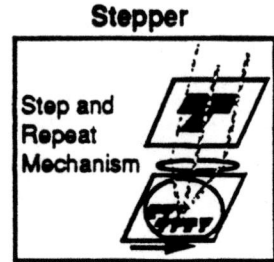

Stepper

Step and
Repeat
Mechanism

STRIPPING: a removal process, usually refers to removal of photoresist from wafer.

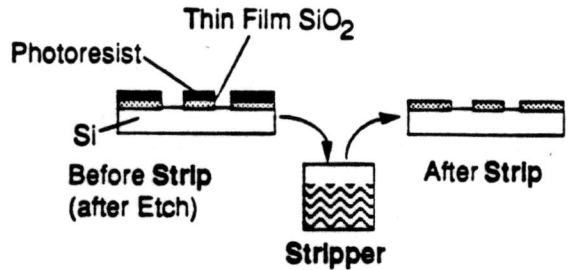

Photoresist

Thin Film SiO$_2$

Si

Before **Strip**
(after Etch)

Stripper

After **Strip**

STYLUS: an arm which is moved along a surface for the purpose of identifying surface variations. A stylus is used to measure film thickness by tracing over a step that is equal to the film depth.

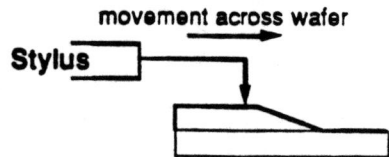

movement across wafer

Stylus

SUBCOLLECTOR: see "Buried Layer".

SUBSTRATE: the underlying material upon which a device, circuit or epitaxial layer is fabricated.

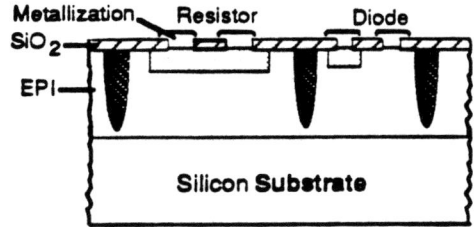

SUBTRACTIVE ETCH: an etch process which results in leaving an island of material on the wafer surface.

SULFURIC ACID (H_2SO_4): a strong acid often used to clean silicon wafers and remove photoresist.

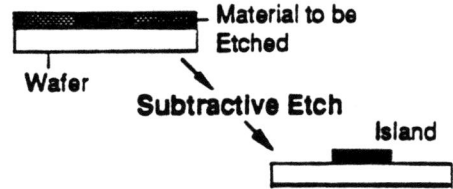

SURFACE STATES: extra donors, acceptors or traps, usually undesired, which may occur on a semiconductor surface because of crystal imperfections or contamination. These surface states may vary undesirably with time.

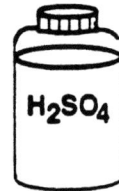

SURFACE INSPECTION: an inspection in which major (gross) physical anomalies in or on a wafer layer are seen through the utilization collimated, white, or U.V. light.

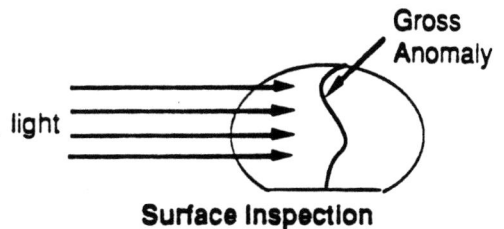

SUSCEPTOR: the flat slab of material (usually graphite) on which wafers are held during high temperature deposition processes such as epitaxial growth or nitride deposition.

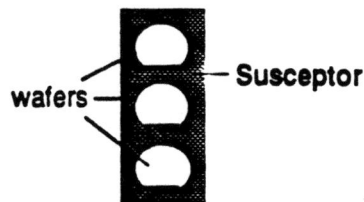

T

TCE: see "Trechloroethylene".

TAPE AUTOMATED BONDING:
process utilizing chip package
leads adhered to a flexible strip
of tape used in automated
bonding of leads to chips.

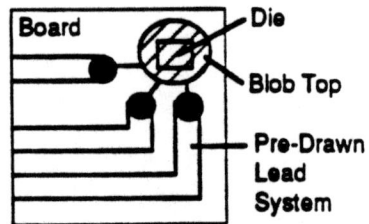

Tape Automated Bonding (TAB)

TARGET: the starting material
bombarded by charged par-
ticles during the sputtering
process. Results are a thin
coating of the target material
deposited on the wafer surface.

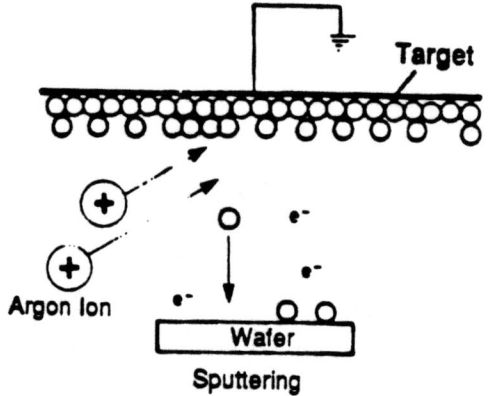

Sputtering

TEST DIE: die on a wafer that
appear to have a different
pattern from most others.
These contain test devices
created by the same processes
as the regular die. The devices
on these die are designed on a
larger scale to allow in-process
quality control.

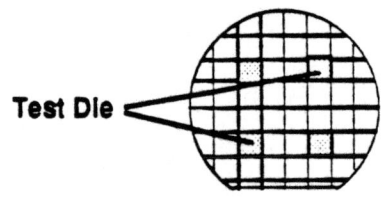

THERMAL ATMOSPHERIC OXIDATION: a process step in which the silicon substrate reacts with oxygen or water vapor to form silicon dioxide in a heated tube furnace at atmospheric pressure.

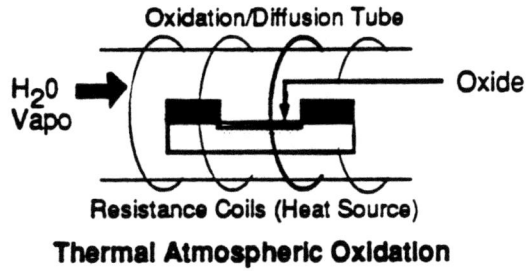

Thermal Atmospheric Oxidation

THERMAL DIFFUSION: a process by which dopant atoms diffuse into the wafer surface by heating the wafer in the range of 1000 C and exposing it to vapors containing the desired dopant.

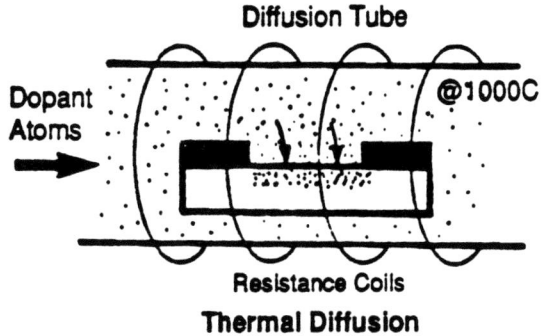

Thermal Diffusion

THERMOCOUPLE: a device used to measure the temperature in a furnace of a reactor. It is made by welding two wires together. Heat generates a voltage between the two materials that is proportional to the temperature.

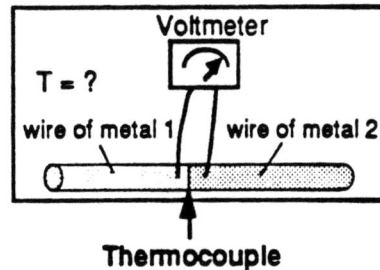

Thermocouple

THIN-FILM INTEGRATED CIRCUIT: a circuit consisting of patterns of tantalum or other materials laid down on a substrate of glass or ceramic, typically larger than silicon integrated circuits. Sometimes designated "FIC."

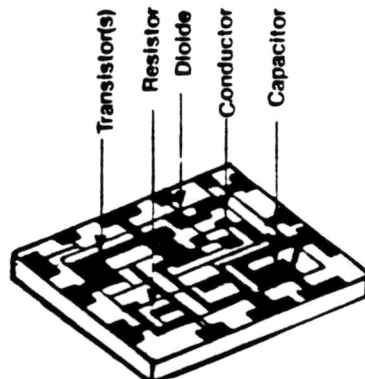

Thin Film Integrated Circuit

THREE FIVE COMPOUND: see "Compound Semiconductor"

THREE POINT PROBE: used to measure electrical characteristics.

TORR: pressure unit. International standard unit replacing the English term, millimeter of mercury (mm of Hg).

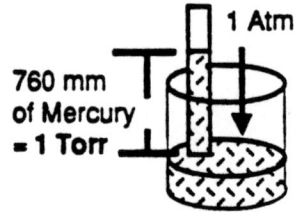

760 mm of Mercury = 1 Torr

1 Atm

TRANSISTOR: a semiconductor device that uses a stream of charge carriers to produce active electronic effects. The name was derived from the electrical characteristic of "transfer resistance." A transistor essentially acts as a current switch and may be a discrete device or a part of an integrated circuit .

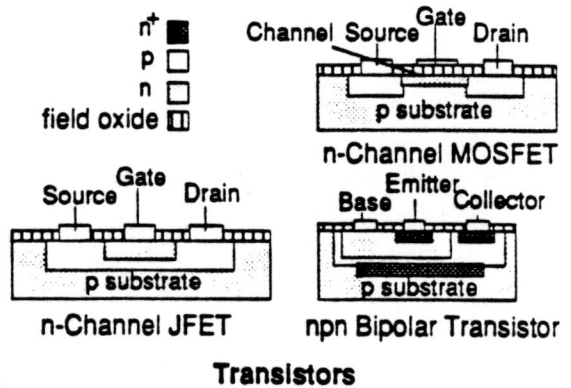

n⁺
p
n
field oxide

Channel Source Gate Drain
p substrate
n-Channel MOSFET

Source Gate Drain
p substrate
n-Channel JFET

Base Emitter Collector
p substrate
npn Bipolar Transistor

Transistors

TRICHLOROETHYLENE (TCE): a solvent used for wafer and general cleaning.

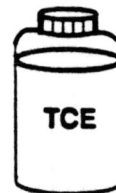

TCE

TUBE: 1. see "Furnace" 2. A cylindrical piece of quartz with fittings on one or both ends. It is placed in a furnace to provide a contamination-free and controlled atmosphere.

Oxidation/Diffusion Tube

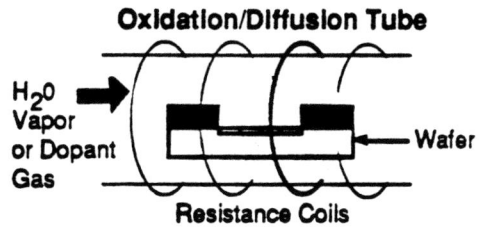

H_2O Vapor or Dopant Gas

Wafer

Resistance Coils

TUNNEL: a concept in clean room design in which each portion of the fab line is isolated. Production equipment lines both sides of the work area (bay) with the manufacturing flow proceeding from station-to-station. Unnecessary portions of the equipment and ancillary equipment are located behind the walls in service areas (chases) that separate the various bays.

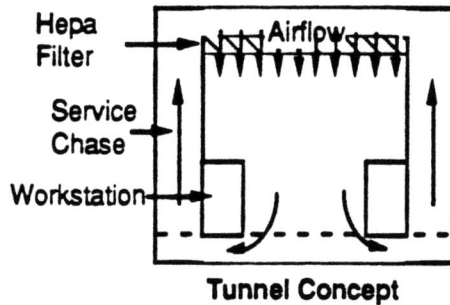

Hepa Filter

Airflow

Service Chase

Workstation

Tunnel Concept

TWINING: a crystal growth defect in which the crystal starts growing in two directions.

Silicon Ingot Exhibiting **Twining**

U

UV: see "Ultraviolet".

ULSI: see "Ultra Large Scale Integration".

UNDERCUTTING: see "Isotropic Etching".

ULTRA LARGE SCALE INTEGRATION (ULSI): Refers to chips with over 1,000,000 components each.

Integration Levels Chart

Level	Abreviation	# Components per Chip
Small Scale Integration	SSI	2 - 50
Medium Scale Integration	MSI	50-5000
Large Scale Integration	LSI	5000-100,000
Very Large Scale Integration	VLSI	100,000-1,000,000
Ultra Large Scale Integration	ULSI	over 1,000,000

ULTRAVIOLET: a light wavelength that is commonly used to cause a response in a photoresist (photosolubilization or photopolymerization).

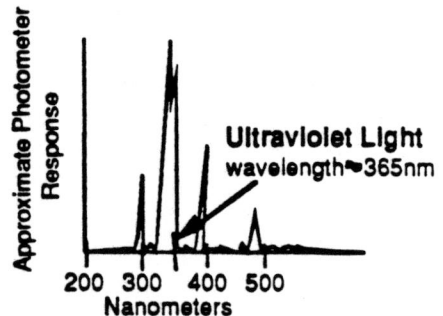

Ultraviolet Light
wavelength≈365nm

UNIPOLAR TRANSISTOR: a transistor such as an FET whose action depends only on majority carriers as compared to bipolar transistors whose operation depends upon both minority and majority carriers.

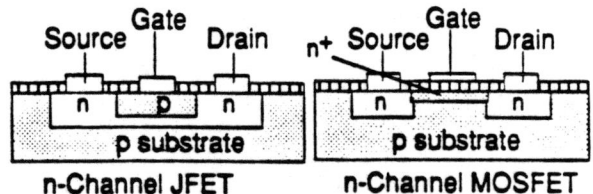

n-Channel JFET   n-Channel MOSFET

Unipolar Tranistors

V

VLSI: see "Very Large Scale Integration".

VACANCY: 1) an empty position in an atomic crystal structure. 2) a type of crystal defect.

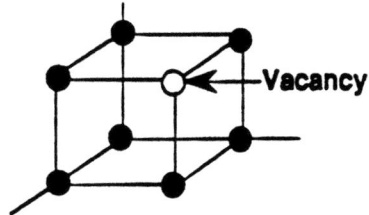

VACUUM: a low pressure condition.

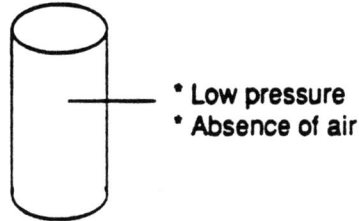

VAPOR PRIMING: a technique in which primer is applied in a vapor state such that the wafer never comes in contact with any possible contamination in the liquid or, in the case of HMDS, any particles of hydrolyzed HMDS.

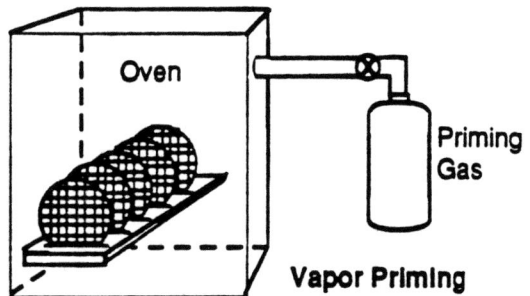

VIA: vertical opening filled with conducting material used to connect circuits on various layers of a device to one another and to the semiconducting substrate. Serves same purpose as "contacts".

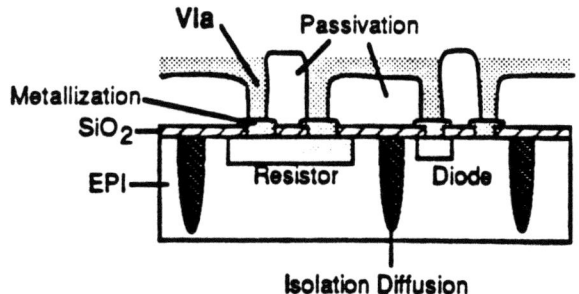

VISCOSITY: the qualitative measure of liquid flow. Viscosity measurements are made by quantifying the force required to move an object through the liquid. It is a measurement of "internal friction".

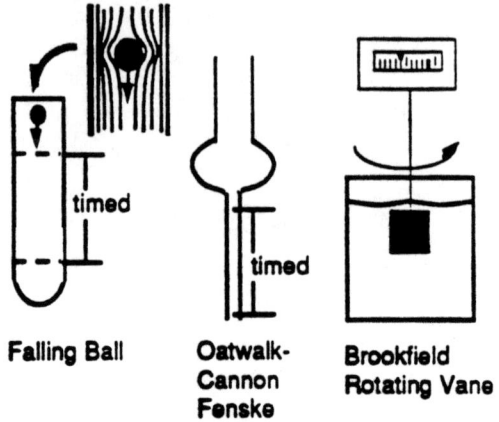

Falling Ball   Oatwalk-Cannon Fenske   Brookfield Rotating Vane

Viscosity Measuring Techniques

VLF HOOD: a work station with vertical laminar air flow designed to keep particulate levels low.

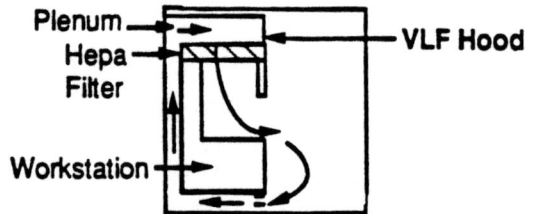

Plenum —— VLF Hood
Hepa Filter
Workstation

VERY LARGE SCALE INTEGRATION (VLSI): Refers to chips with between 100,000 and 1,000,000 components.

Integration Levels Chart

Level	Abreviation	# Components per Chip
Small Scale Integration	SSI	2 - 50
Medium Scale Integration	MSI	50-5000
Large Scale Integration	LSI	5000-100,000
Very Large Scale Integration	VLSI	100,000-1,000,000
Ultra Large Scale Integration	ULSI	over 1,000,000

VMOS: V-groove MOS technology. A transistor which requires V-shaped grooves in the wafer surface. Components are formed on the sides of the grooves. Density is increased as a consequence of the greater surface area and problems associated with surface contamination are eliminated.

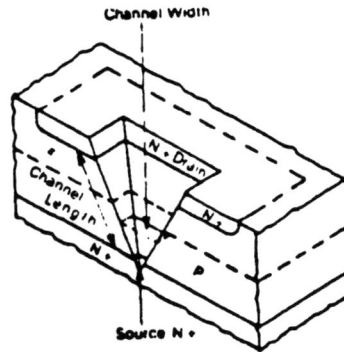

VMOS Structure

VOLTAGE: the force applied between two points causing charged particles (electrical current) to flow, measured in units of "volts".

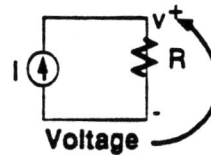

Voltage

W

WAFER: a thin, usually round, slice of a semiconductor material from which chips are made.

Wafer

WAFER CODING: refers to grinding straight edges in ingots out of which wafer slices are made. Done to identify type of crystal orientation and dopant type as dictated by SEMI standards. Wafers, which are circular in shape, have at least one straight edge called a "wafer flat".

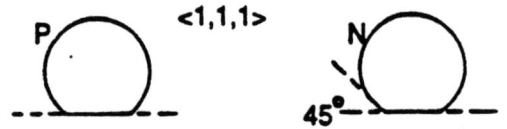

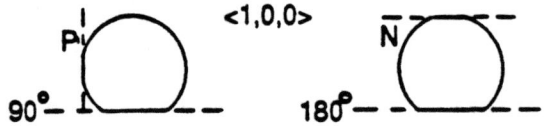

P   <1,1,1>   N   45°

P   <1,0,0>   N

90°   180°

Wafer Coding

WAFER FAB: site where integrated circuits or devices are manufactured.

WAFER FABRICATION: process of manufacturing wafers

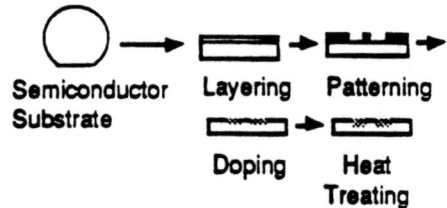

Semiconductor Substrate   Layering   Patterning

Doping   Heat Treating

Wafer Fabrication

WAFER FLAT: flat area(s) ground onto the wafers edges to indicate the crystal orientation of the wafer material and the dopant type.

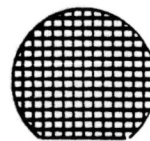

Wafer Flat

110

WAFER SCRUBBER: a mechanical scrubber which holds the wafer on a rotating chuck with a vacuum. A rotating fiber brush and/or a stream of pressurized water is brought in contact with the rotating wafer.

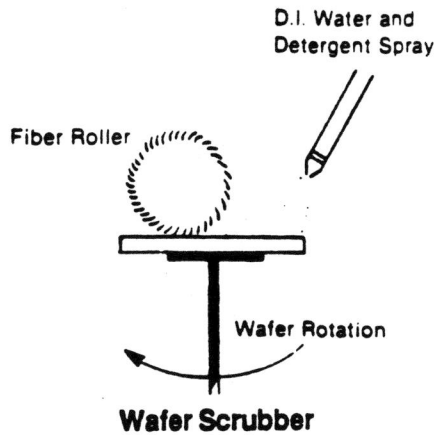

D.I. Water and Detergent Spray

Fiber Roller

Wafer Rotation

Wafer Scrubber

WAFER SORT: the step after wafer fabrication during which the electrical parameters of integrated circuits are tested for functionality. Probes contact the pads of the circuit to conduct the test leading to the name "prober" for the equipment that performs electrical tests on each die site of completed wafers.

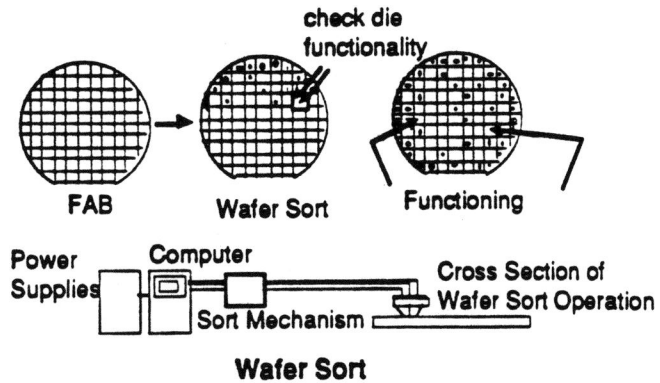

check die functionality

FAB   Wafer Sort   Functioning

Power Supplies   Computer

Sort Mechanism

Cross Section of Wafer Sort Operation

Wafer Sort

WAFER SORT YIELD: the number of functioning die at wafer sort as compared to the total number of die started. Typically, the lowest major yield point for integrated circuits.

Wafer Sort Yield =
$$\frac{\text{\# functioning die at wafer sort}}{\text{\# die started}} \times 100\%$$

WHITE ELEPHANT: see "Elephant"

WIRE BONDING: an assembly step in which thin conductive wires are attached between the die bonding pads and the lead connections in the package.

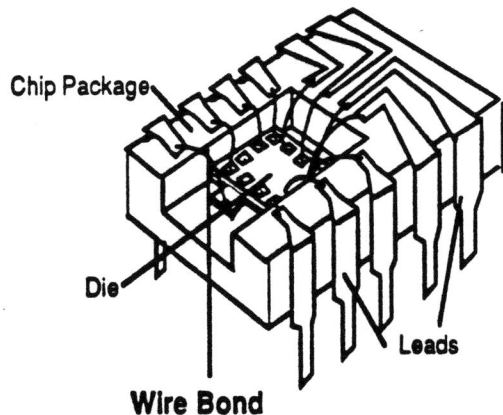

Chip Package

Die

Leads

Wire Bond

111

XYZ

X-RAY EXPOSURE SYSTEM: imaging system using X-rays as the exposure source. Due to their short wavelengths, x-rays exhibit no detrimental diffraction effects.

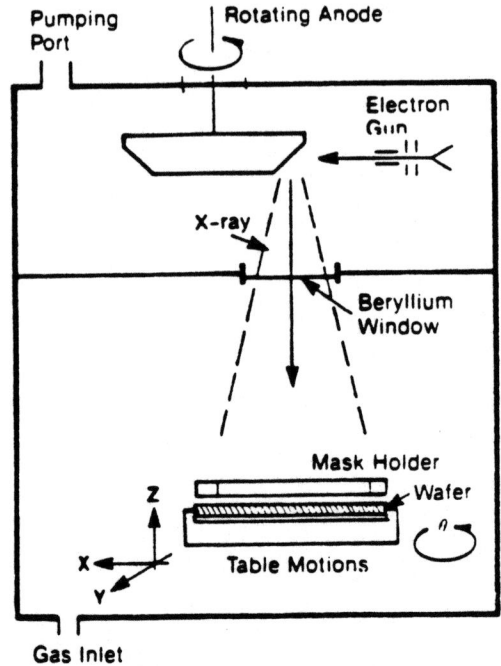

XYLENE: a primer chemical which dehydrates the surface of the wafer to promote better adhesion. Also used as a developer for negative photoresist.

YELLOW ROOM: areas in wafer manufacturing plant (fab) where yellow filters are used to block out ultraviolet light which can adversely affect photoresist.

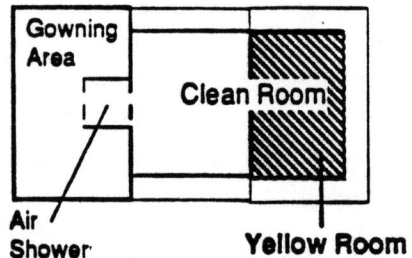

YIELD: a percentage used in the semiconductor industry which indicates the amount of finished products leaving a process as compared to the amount of product entering that process.

$$\text{Yield} = \frac{\text{\# good die out of process}}{\text{\# die started into process}} \times 100\%$$

YIELD BUST: very low or no yield.

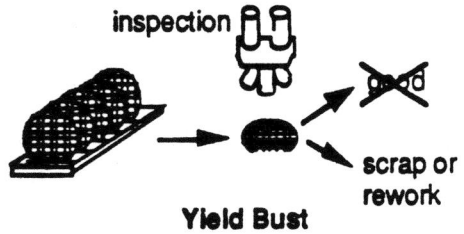

inspection

scrap or rework

Yield Bust

ℓ